MPEG-2

For Howard and Matthew

MPEG-2

John Watkinson

Focal Press

OXFORD   AUCKLAND   BOSTON   JOHANNESBURG   MELBOURNE   NEW DELHI

Focal Press
An imprint of Butterworth-Heinemann
Linacre House, Jordan Hill, Oxford OX2 8DP
225 Wildwood Avenue, Woburn, MA 01801-2041
A division of Reed Educational and Professional Publishing Ltd

ℛ A member of the Reed Elsevier plc group

First published 1999

British Library Cataloguing in Publication Data
A catalogue record for this book is available from the British Library

Library of Congress Cataloguing in Publication Data
A catalogue record for this book is available from the Library of Congress

ISBN 0 240 51510 2

Composition by Genesis Typesetting, Rochester, Kent
Printed and bound in Great Britain

Contents

Preface

Digital compression technology has been employed for a long time, but until recently the technology was too complex for everyday applications. The onward march of LSI technology means that increasingly complex processes become available at moderate cost. Digital compression has now reached the stage where it can economically be applied to video and audio systems on a wide scale. MPEG-2 has assisted in the process by providing interoperability standards for audio and video compression and communication.

Economic compression does more than make existing processes cheaper. It allows new developments which were impossible without it. This book recognizes the wide application of MPEG by treating the subject from first principles without assuming any particular background for the reader. MPEG is not the only compression technique and other approaches will be mentioned here for completeness. While compression is traditionally described mathematically, it is the author's view that mathematics is no more than a technical form of shorthand *describing* a process. As such it cannot *explain* anything and has no place in a book of this kind where instead the explanations are in plain English.

An introductory chapter is included which suggests some applications of MPEG and how it works in a simplified form. Additionally a fundamentals chapter contains all the background necessary to follow the rest of the book.

Compression theory is balanced with a great deal of practice. The creation of MPEG Elementary Streams and their multiplexing into transport streams is dealt with in detail along with the problems involved in synchronizing all the signals in a multiplex. The practical testing of MPEG transport streams is also considered.

Throughout this book the reader will find notes of caution and outlines of various pitfalls for the unwary. Compression is a useful tool in certain applications but it is a dangerous master and if used indiscriminately the results are bound to be disappointing. Unfortunately there are signs that this is happening. Various descriptions of kinds of artifacts and impairments which result from the misuse of compression are included here. Compression is rather like a box of fireworks; used wisely a pleasing display results – used carelessly it can blow up in your face.

To suggest that compression should never be used to ensure that disasters are avoided is as naive as trying to ban fireworks. The solution is the same: to explain the dangers and propose safe practices. Retire to a safe place and read this book.

John Watkinson
Burghfield Common 1999

Acknowledgements

Information for this book has come from a range of sources to whom I am indebted. I would specifically mention the publications of the ISO, the AES and SMPTE which provided essential groundwork. I have had numerous discussions with Peter de With of Philips, Steve Lyman of CBC, Bruce Devlin and Mike Knee of Snell and Wilcox and Peter Kraniauskas which have all been extremely useful. The assistance of MicroSoft Corp. and Tektronix Inc. is also appreciated.

1

Introduction to compression

1.1   What is MPEG-2?

MPEG is actually an acronym for the moving pictures experts group which was formed by the ISO (International Standards Organization) to set standards for audio and video compression and transmission. The first compression standard for audio and video was MPEG-1[1,2] but this was of limited application and the subsequent MPEG-2 standard was considerably broader in scope and of wider appeal.[3] For example, MPEG-2 supports interlace whereas MPEG-1 did not.

Compression is summarized in Figure 1.1. It will be seen in (a) that the data rate is reduced at source by the *compressor*. The compressed data are then passed through a communication channel and returned to the original rate by the *expander*. The ratio between the source data rate and the channel data rate is called the *compression factor*. The term *coding gain* is also used. Sometimes a compressor and expander in series are referred to as a *compander*. The compressor may equally well be referred to as a *coder* and the expander a *decoder* in which case the tandem pair may be called a *codec*.

In audio and video compression, where the encoder is more complex than the decoder the system is said to be asymmetrical. Figure 1.1(b) shows that MPEG works in this way. The encoder needs to be algorithmic or adaptive whereas the decoder is 'dumb' and carries out fixed actions. This is advantageous in applications such as broadcasting where the number of expensive complex encoders is small but the number of simple inexpensive decoders is large. In point-to-point applications the advantage of asymmetrical coding is not so great.

The approach of the ISO to standardization in MPEG is novel because it is not the encoder which is standardized. Figure 1.2(a) shows that instead the way in which a decoder shall interpret the bitstream is

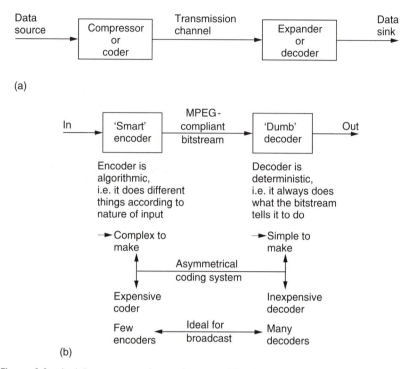

Figure 1.1 In (a) a compression system consists of compressor or coder, a transmission channel and a matching expander or decoder. The combination of coder and decoder is known as a codec. (b) MPEG is asymmetrical since the encoder is much more complex than the decoder.

defined. A decoder which can successfully interpret the bitstream is said to be *compliant*. Figure 1.2(b) shows that the advantage of standardizing the decoder is that over time encoding algorithms can improve yet compliant decoders will continue to function with them.

Manufacturers can supply encoders using algorithms which are proprietary and their details do not need to be published. A useful result is that there can be competition between different encoder designs which means that better designs will evolve. The user will have greater choice because different levels of cost and complexity can exist in a range of coders yet a compliant decoder will operate with them all.

MPEG is, however, much more than a compression scheme as it also standardizes the protocol and syntax under which it is possible to combine or multiplex audio data with video data to produce a digital equivalent of a television program. Many such programs can be combined in a single multiplex and MPEG defines the way in which such multiplexes can be created and transported. The definitions include the metadata which decoders require to demultiplex correctly and which users will need to locate programs of interest.

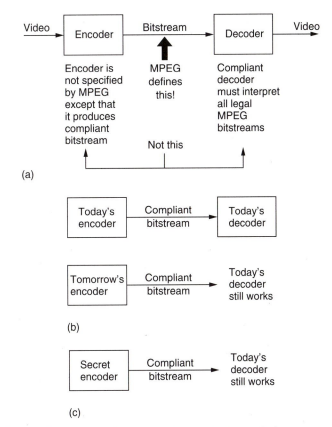

Figure 1.2 (a) MPEG defines the protocol of the bitstream between encoder and decoder. The decoder is defined by implication, the encoder is left very much to the designer. (b) This approach allows future encoders of better performance to remain compatible with existing decoders. (c) This approach also allows an encoder to produce a standard bitstream while its technical operation remains a commercial secret.

As with all video systems there is a requirement for synchronizing or genlocking and this is particularly complex when a multiplex is assembled from many signals which are not necessarily synchronized to one another.

1.2 Why compression is necessary

Compression, bit rate reduction and data reduction are all terms which mean basically the same thing in this context. In essence the same (or nearly the same) information is carried using a smaller quantity or rate of data. It should be pointed out that, in audio, *compression* traditionally means a process in which the dynamic range of the sound is reduced. In the context of MPEG the same word means that the bit rate is reduced, ideally leaving the dynamics of the signal unchanged. Provided the context is clear, the two meanings can co-exist without a great deal of confusion.

There are several reasons why compression techniques are popular:

1. Compression extends the playing time of a given storage device.
2. Compression allows miniaturization. With less data to store, the same playing time is obtained with smaller hardware. This is useful in ENG (electronic news gathering) and consumer devices.
3. Tolerances can be relaxed. With less data to record, storage density can be reduced making equipment which is more resistant to adverse environments and which requires less maintenance.
4. In transmission systems, compression allows a reduction in bandwidth which will generally result in a reduction in cost. It may make possible some process which would be impracticable without it.
5. If a given bandwidth is available to an uncompressed signal, compression allows faster than real-time transmission in the same bandwidth.
6. If a given bandwidth is available, compression allows a better-quality signal in the same bandwidth.

1.3   Some applications of MPEG-2

The applications of audio and video compression are limitless and the ISO has done well to provide standards which are appropriate to the wide range of possible compression products. MPEG-2 embraces video pictures from the tiny screen of a videophone to the high-definition images needed for electronic cinema. Audio coding stretches from speech grade mono to multichannel surround sound.

Figure 1.3 shows the use of a codec with a recorder. The playing time of the medium is extended in proportion to the compression factor. In the case of tapes, the access time is improved because the length of tape needed for a given recording is reduced and so it can be rewound more quickly.

In the case of DVD (Digital Video Disk aka Digital Versatile Disk) the challenge was to store an entire movie on one 12 cm disk. The storage density available with today's optical disk technology is such that recording of conventional uncompressed video would be out of the question.

In communications, the cost of data links is often roughly proportional to the data rate and so there is simple economic pressure to use a high compression factor. However, it should be borne in mind that implementing the codec also has a cost which rises with compression factor and so a degree of compromise will be inevitable.

In the case of Video-on-Demand, technology exists to convey full bandwidth video to the home, but to do so for a single individual at the moment would be prohibitively expensive. Without compression, HDTV (high-definition television) requires too much bandwidth. With compression, HDTV can be transmitted to the home in a similar bandwidth to an existing analog SDTV channel. Compression does not make Video-on-Demand or HDTV possible, it makes them economically viable.

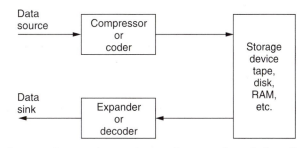

Figure 1.3 Compression can be used around a recording medium. The storage capacity may be increased or the access time reduced according to the application.

In workstations designed for the editing of audio and/or video, the source material is stored on hard disks for rapid access. While top-grade systems may function without compression, many systems use compression to offset the high cost of disk storage. When a workstation is used for *off-line* editing, a high compression factor can be used and artifacts will be visible in the picture.

This is of no consequence as the picture is only seen by the editor who uses it to make an EDL (Edit Decision List) which is no more than a list of actions and the timecodes at which they occur. The original uncompressed material is then *conformed* to the EDL to obtain a high-quality edited work. When *on-line* editing is being performed, the output of the workstation is the finished product and clearly a lower compression factor will have to be used.

Perhaps it is in broadcasting where the use of compression will have its greatest impact. There is only one electromagnetic spectrum and pressure from other services such as cellular telephones makes efficient use of bandwidth mandatory. Analog television broadcasting is an old technology and makes very inefficient use of bandwidth. Its replacement by a compressed digital transmission will be inevitable for the practical reason that the bandwidth is needed elsewhere.

Fortunately, in broadcasting there is a mass market for decoders and these can be implemented as low-cost integrated circuits. Fewer encoders are needed and so it is less important if these are expensive. While the cost of digital storage goes down year on year, the cost of electromagnetic spectrum goes up. Consequently in the future the pressure to use compression in recording will ease whereas the pressure to use it in radio communications will increase.

1.4   Lossless and perceptive coding

Although there are many different coding techniques, all of them fall into one or other of these categories. In *lossless* coding, the data from the expander are identical bit-for-bit with the original source data. The so-called 'stacker' programs which increase the apparent capacity of disk drives in personal computers use lossless codecs. Clearly with computer programs the corruption of a single bit can be catastrophic. Lossless coding is generally restricted to compression factors of around 2:1.

It is important to appreciate that a lossless coder cannot guarantee a particular compression factor and the communications link or recorder used with it must be able to function with the variable output data rate. Source data which result in poor compression factors on a given codec are described as *difficult*. It should be pointed out that the difficulty is often a function of the codec. In other words, data which one codec finds difficult may not be found difficult by another. Lossless codecs can be included in bit-error-rate testing schemes. It is also possible to cascade or *concatenate* lossless codecs without any special precautions.

In *lossy* coding data from the expander are not identical bit-for-bit with the source data and as a result comparing the input with the output is bound to reveal differences. Lossy codecs are not suitable for computer data, but are used in MPEG as they allow greater compression factors than lossless codecs. Successful lossy codecs are those in which the errors are arranged so that a human viewer or listener finds them subjectively difficult to detect. Thus lossy codecs must be based on an understanding of psychoacoustic and psychovisual perception and are often called *perceptive* codes.

In perceptive coding, the greater the compression factor required, the more accurately must the human senses be modelled. Perceptive coders can be forced to operate at a fixed compression factor. This is convenient for practical transmission applications where a fixed data rate is easier to handle than a variable rate. The result of a fixed compression factor is that the subjective quality can vary with the 'difficulty' of the input material. Perceptive codecs should not be concatenated indiscriminately especially if they use different algorithms. As the reconstructed signal from a perceptive codec is not bit-for-bit accurate, clearly such a codec cannot be included in any bit-error-rate testing system as the coding differences would be indistinguishable from real errors.

Although the adoption of digital techniques is recent, compression itself is as old as television. Figure 1.4 shows some of the compression techniques used in traditional television systems.

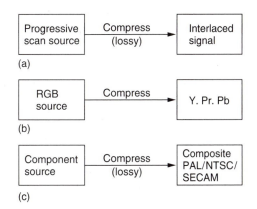

Figure 1.4 Compression is as old as television. (a) Interlace is a primitive way of halving the bandwidth. (b) Colour difference working invisibly reduces colour resolution. (c) Composite video transmits colour in the same bandwidth as monochrome.

One of the oldest techniques is interlace which has been used in analog television from the very beginning as a primitive way of reducing bandwidth. As will be seen in Chapter 5, interlace is not without its problems, particularly in motion rendering. MPEG-2 supports interlace simply because legacy interlaced signals exist and there is a requirement to compress them. This should not be taken to mean that it is a good idea.

The generation of colour difference signals from RGB in video represents an application of perceptive coding. The human visual system (HVS) sees no change in quality although the bandwidth of the colour difference signals is reduced. This is because human perception of detail in colour changes is much less than in brightness changes. This approach is sensibly retained in MPEG.

Composite video systems such as PAL, NTSC and SECAM are all analog compression schemes which embed a subcarrier in the luminance signal so that colour pictures are available in the same bandwidth as monochrome. In comparison with a progressive scan RGB picture, interlaced composite video has a compression factor of 6:1.

In a sense MPEG-2 can be considered to be a modern digital equivalent of analog composite video as it has most of the same attributes. For example, the eight-field sequence of PAL subcarrier which makes editing diffficult has its equivalent in the GOP (group of pictures) of MPEG.

1.5 Compression principles

In a PCM digital system the bit rate is the product of the sampling rate and the number of bits in each sample and this is generally constant. Nevertheless the *information* rate of a real signal varies. In all real signals, part of the signal is obvious from what has gone before or what may come later and a suitable receiver can predict that part so that only the true information actually has to be sent. If the characteristics of a predicting receiver are known, the transmitter can omit parts of the message in the knowledge that the receiver has the ability to recreate it. Thus all encoders must contain a model of the decoder.

One definition of information is that it is the unpredictable or surprising element of data. Newspapers are a good example of information because they only mention items which are surprising. Newspapers never carry items about individuals who have *not* been involved in an accident as this is the normal case. Consequently the phrase 'no news is good news' is remarkably true because if an information channel exists but nothing has been sent then it is most likely that nothing remarkable has happened.

The unpredictability of the punch line is a useful measure of how funny a joke is. Often the build-up paints a certain picture in the listener's imagination, which the punch line destroys utterly. One of the author's favourites is the one about the newly married couple who didn't know the difference between putty and petroleum jelly – their windows fell out.

The difference between the information rate and the overall bit rate is known as the redundancy. Compression systems are designed to eliminate as much of that redundancy as practicable or perhaps affordable. One way in which this can be done is to exploit statistical predictability in signals. The information content or *entropy* of a sample is a function of how different it is from the predicted value. Most signals have some degree of predictability. A sine wave is highly predictable, because all cycles look the same. According to Shannon's theory, any signal which is totally predictable carries no information. In the case of the sine wave this is clear because it represents a single frequency and so has no bandwidth.

At the opposite extreme a signal such as noise is completely unpredictable and as a result all codecs find noise *difficult*. There are two consequences of this characteristic. First, a codec which is designed using the statistics of real material should not be tested with random noise

eliminate the same predictable degree signal (handwritten margin note)

because it is not a representative test. Second, a codec which performs well with clean source material may perform badly with source material containing superimposed noise. Most practical compression units require some form of pre-processing before the compression stage proper and appropriate noise reduction should be incorporated into the pre-processing if noisy signals are anticipated. It will also be necessary to restrict the degree of compression applied to noisy signals.

All real signals fall part way between the extremes of total predictability and total unpredictability or noisiness. If the bandwidth (set by the sampling rate) and the dynamic range (set by the wordlength) of the transmission system are used to delineate an area, this sets a limit on the information capacity of the system. Figure 1.5(a) shows that most real signals only occupy part of that area. The signal may not contain all frequencies, or it may not have full dynamics at certain frequencies.

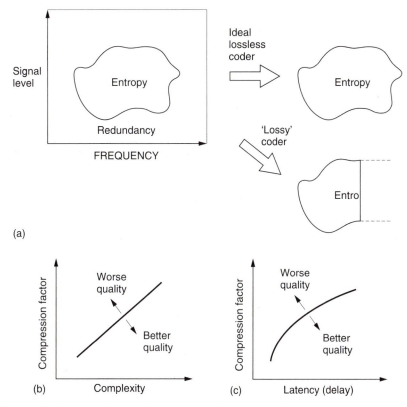

Figure 1.5 (a) A perfect coder removes only the redundancy from the input signal and results in subjectively lossless coding. If the remaining entropy is beyond the capacity of the channel some of it must be lost and the codec will then be lossy. An imperfect coder will also be lossy as it fails to keep all entropy. (b) As the compression factor rises, the complexity must also rise to maintain quality. (c) High compression factors also tend to increase latency or delay through the system.

Entropy can be thought of as a measure of the actual area occupied by the signal. This is the area that *must* be transmitted if there are to be no subjective differences or *artifacts* in the received signal. The remaining area is called the *redundancy* because it adds nothing to the information conveyed. Thus an ideal coder could be imagined which miraculously sorts out the entropy from the redundancy and only sends the former. An ideal decoder would then recreate the original impression of the information quite perfectly.

As the ideal is approached, the coder complexity and the latency or delay both rise. Figure 1.5(b) shows how complexity increases with compression factor and (c) shows how increasing the codec latency can improve the compression factor. Obviously we would have to provide a channel which could accept whatever entropy the coder extracts in order to have transparent quality. As a result moderate coding gains which only remove redundancy need not cause artifacts and result in systems which are described as *subjectively lossless*.

If the channel capacity is not sufficient for that, then the coder will have to discard some of the entropy and with it useful information. Larger coding gains which remove some of the entropy must result in artifacts. It will also be seen from Figure 1.5 that an imperfect coder will fail to separate the redundancy and may discard entropy instead, resulting in artifacts at a sub-optimal compression factor.

A single variable-rate transmission or recording channel is inconvenient and unpopular with channel providers because it is difficult to police. The requirement can be overcome by combining several compressed channels into one constant rate transmission in a way which flexibly allocates data rate between the channels. Provided the material is unrelated, the probability of all channels reaching peak entropy at once is very small and so those channels which are at one instant passing easy material will free up transmission capacity for those channels which are handling difficult material. This is the principle of statistical multiplexing.

Where the same type of source material is used consistently, e.g. English text, then it is possible to perform a statistical analysis on the frequency with which particular letters are used. Variable-length coding is used in which frequently used letters are allocated short codes and letters which occur infrequently are allocated long codes. This results in a lossless code. The well-known Morse code used for telegraphy is an example of this approach. The letter e is the most frequent in English and is sent with a single dot.

An infrequent letter such as z is allocated a long complex pattern. It should be clear that codes of this kind which rely on a prior knowledge of the statistics of the signal are only effective with signals actually having those statistics. If Morse code is used with another language, the transmission becomes significantly less efficient because the statistics are quite different; the letter z, for example, is quite common in Czech.

The Huffman code[4] is one which is designed for use with a data source having known statistics and shares the same principles with the Morse code. The probability of the different code values to be transmitted is studied, and the most frequent codes are arranged to be transmitted with short wordlength symbols. As the probability of a code value falls, it will be allocated longer wordlength. The Huffman code is used in conjunction with a number of compression techniques and is shown in Figure 1.6.

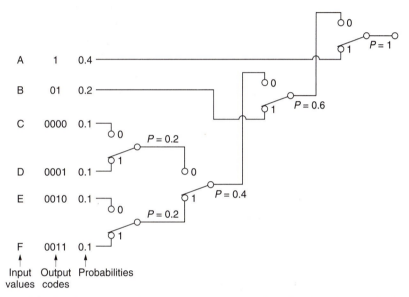

Figure 1.6 The Huffman code achieves compression by allocating short codes to frequent values. To aid descrializing the short codes are not prefixes of longer codes.

The input or *source* codes are assembled in order of descending probability. The two lowest probabilities are distinguished by a single code bit and their probabilities are combined. The process of combining probabilities is continued until unity is reached and at each stage a bit is used to distinguish the path. The bit will be a zero for the most probable path and one for the least. The compressed output is obtained by reading the bits which describe which path to take going from right to left.

In the case of computer data, there is no control over the data statistics. Data to be recorded could be instructions, images, tables, text files and so on; each having their own code value distributions. In this case a coder relying on fixed source statistics will be completely inadequate. Instead a system is used which can learn the statistics as it goes along. The Lempel–Ziv–Welch (LZW) lossless codes are in this category. These codes build up a conversion table between frequent long source data strings and short transmitted data codes at both coder and decoder and initially their compression factor is below unity as the contents of the conversion tables are transmitted along with the data. However, once the tables are established, the coding gain more than compensates for the initial loss. In some applications, a continuous analysis of the frequency of code selection is made and if a data string in the table is no longer being used with sufficient frequency it can be deselected and a more common string substituted.

Lossless codes are less common for audio and video coding where perceptive codes are permissible. The perceptive codes often obtain a coding gain by shortening the wordlength of the data representing the signal waveform. This must increase the noise level and the trick is to ensure that the resultant noise is placed at frequencies where human senses are least able to perceive it. As a result although the received signal is measurably different from the source data, it can *appear* the same to the human listener or viewer at moderate compression factors. As these codes rely on the characteristics of human sight and hearing, they can only be fully tested subjectively.

The compression factor of such codes can be set at will by choosing the wordlength of the compressed data. While mild compression will be undetectible, with greater compression factors, artefacts become noticeable. Figure 1.5 shows that this is inevitable from entropy considerations.

1.6   Audio compression

Perceptive coding in audio relies on the principle of auditory masking, which is treated in detail in section 4.1. Masking causes the ear/brain combination to be less sensitive to sound at one frequency in the presence of another at a nearby frequency. If a first tone is present in the input, then it will mask signals of lower level at nearby frequencies. The quantizing of the first tone and of further tones at those frequencies can be made coarser. Fewer bits are needed and a coding gain results. The increased quantizing error is allowable if it is masked by the presence of the first tone.

1.6.1   Sub-band coding

Sub-band coding mimics the frequency analysis mechanism of the ear and splits the audio spectrum into a large number of different bands. Signals in these bands can then be quantized independently. The quantizing error which results is confined to the frequency limits of the band and so it can be arranged to be masked by the program material. The techniques used in Layers 1 and 2 of MPEG audio are based on sub-band coding as are those used in DCC (Digital Compact Cassette).

1.6.2 Transform coding

In transform coding the time-domain audio waveform is converted into a frequency domain representation such as a Fourier, discrete cosine or wavelet transform (see Chapter 3). Transform coding takes advantage of the fact that the amplitude or envelope of an audio signal changes relatively slowly and so the coefficients of the transform can be transmitted relatively infrequently. Clearly such an approach breaks down in the presence of transients and adaptive systems are required in practice. Transients cause the coefficients to be updated frequently whereas in stationary parts of the signal such as sustained notes the update rate can be reduced. Discrete cosine transform (DCT) coding is used in Layer III of MPEG audio and in the compression system of the Sony MiniDisc.

1.6.3 Predictive coding

In a predictive coder there are two identical predictors, one in the coder and one in the decoder. Their job is to examine a run of previous sample code values and to extrapolate forward to estimate or predict what the next code value will be. This is subtracted from the *actual* next code value at the encoder to produce a prediction error which is transmitted. The decoder then adds the prediction error to its own prediction to obtain the output code value again. Predictive coders work with a short encode and decode delay and are useful in telephony where a long loop delay causes problems.

1.7 Video compression

Video signals exist in four dimensions: these are the attributes of the sample, the horizontal and vertical spatial axes and the time axis. Compression can be applied in any or all of those four dimensions. MPEG-2 assumes 8-bit colour difference signals as the input, requiring rounding if the source is 10-bit. The sampling rate of the colour signals is less than that of the luminance. This is done by downsampling the colour samples horizontally and generally vertically as well. Essentially an

MPEG-2 system has three parallel simultaneous channels, one for luminance and two colour difference, which after coding are multiplexed into a single bitstream.

Figure 1.7(a) shows that when individual pictures are compressed without reference to any other pictures, the time axis does not enter the process which is therefore described as *intra-coded* (intra = within) compression. The term *spatial coding* will also be found. It is an advantage of intra-coded video that there is no restriction to the editing which can be carried out on the picture sequence. As a result compressed VTRs such as Digital Betacam, DVC and D-9 use spatial coding. Cut editing may take place on the compressed data directly if necessary. As spatial coding treats each picture independently, it can employ certain techniques developed for the compression of still pictures. The ISO JPEG (Joint Photographic Experts Group) compression standards[5,6] are in this category. Where a succession of JPEG coded images are used for television, the term 'Motion JPEG' will be found.

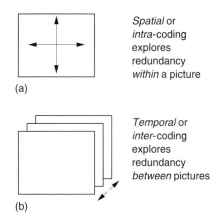

Spatial or *intra*-coding explores redundancy *within* a picture

(a)

Temporal or *inter*-coding explores redundancy *between* pictures

(b)

Figure 1.7 (a) Spatial or intra-coding works on individual images. (b) Temporal or inter-coding works on successive images.

Greater compression factors can be obtained by taking account of the redundancy from one picture to the next. This involves the time axis, as Figure 1.7(b) shows, and the process is known as *inter-coded* (inter = between) or *temporal* compression.

Temporal coding allows a higher compression factor, but has the disadvantage that an individual picture may exist only in terms of the differences from a previous picture. Clearly editing must be undertaken with caution and arbitrary cuts simply cannot be performed on the MPEG bitstream. If a previous picture is removed by an edit, the difference data will then be insufficient to recreate the current picture.

1.7.1 Intra-coded compression

Intra-coding works in three dimensions on the horizontal and vertical spatial axes and on the sample values. Analysis of typical television pictures reveals that while there is a high spatial frequency content due to detailed areas of the picture, there is a relatively small amount of energy at such frequencies. Often pictures contain sizeable areas in which the same or similar pixel values exist. This gives rise to low spatial frequencies. The average brightness of the picture results in a substantial zero-frequency component. Simply omitting the high-frequency components is unacceptable as this causes an obvious softening of the picture.

A coding gain can be obtained by taking advantage of the fact that the amplitude of the spatial components falls with frequency. It is also possible to take advantage of the eye's reduced sensitivity to noise in high spatial frequencies. If the spatial frequency spectrum is divided into frequency bands the high-frequency bands can be described by fewer bits not only because their amplitudes are smaller but also because more noise can be tolerated. The wavelet transform and the discrete cosine transform used in MPEG allows two-dimensional pictures to be described in the frequency domain and these are discussed in Chapter 3.

1.7.2 Inter-coded compression

Inter-coding takes further advantage of the similarities between successive pictures in real material. Instead of sending information for each picture separately, inter-coders will send the difference between the previous picture and the current picture in a form of differential coding. Figure 1.8 shows the principle. A picture store is required at the coder to allow comparison to be made between successive pictures and a similar store is required at the decoder to make the previous picture available. The difference data may be treated as a picture itself and subjected to some form of transform-based spatial compression.

The simple system of Figure 1.8(a) is of limited use as in the case of a transmission error, every subsequent picture would be affected. Channel switching in a television set would also be impossible. In practical systems a modification is required. One approach is the so-called 'leaky predictor' in which the next picture is predicted from a limited number of previous pictures rather than from an indefinite number. As a result errors cannot propagate indefinitely. The approach used in MPEG is that periodically some absolute picture data are transmitted in place of difference data.

Figure 1.8(b) shows that absolute picture data, known as *I* or *intra pictures* are interleaved with pictures which are created using difference

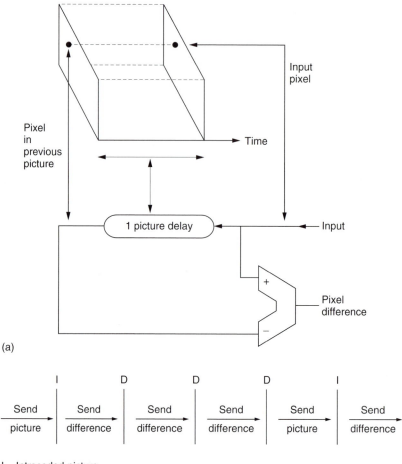

(a)

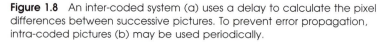

I = Intracoded-picture

D = Differentially coded picture

(b)

Figure 1.8   An inter-coded system (a) uses a delay to calculate the pixel differences between successive pictures. To prevent error propagation, intra-coded pictures (b) may be used periodically.

data, known as *P* or *predicted* pictures. The *I* pictures require a large amount of data, whereas the *P* pictures require less data. As a result the instantaneous data rate varies dramatically and buffering has to be used to allow a constant transmission rate. The leaky predictor needs less buffering as the compression factor does not change so much from picture to picture.

The *I* picture and all the *P* pictures prior to the next *I* picture are called a group of pictures (GOP). For a high compression factor, a large number of *P* pictures should be present between *I* pictures, making a long GOP.

However, a long GOP delays recovery from a transmission error. The compressed bitstream can only be edited at *I* pictures as shown.

In the case of moving objects, although their appearance may not change greatly from picture to picture, the data representing them on a fixed sampling grid will change and so large differences will be generated between successive pictures. It is a great advantage if the effect of motion can be removed from difference data so that they only reflect the changes in appearance of a moving object since a much greater coding gain can then be obtained. This is the objective of motion compensation.

1.7.3   Introduction to motion compensation

In real television program material objects move around before a fixed camera or the camera itself moves. Motion compensation is a process which effectively measures motion of objects from one picture to the next so that it can allow for that motion when looking for redundancy between pictures. Figure 1.9 shows that moving pictures can be expressed in a three-dimensional space which results from the screen area moving along the time axis. In the case of still objects, the only motion is along the time axis. However, when an object moves, it does so along the *optic flow axis* which is not parallel to the time axis. The optic flow axis joins the same point on a moving object as it takes on various screen positions.

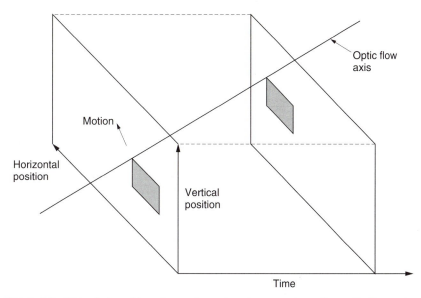

Figure 1.9 Objects travel in a three-dimensional space along the optic flow axis which is only parallel to the time axis if there is no movement.

It will be clear that the data values representing a moving object change with respect to the time axis. However, looking along the optic flow axis the appearance of an object only changes if it deforms, moves into shadow or rotates. For simple translational motions the data representing an object are highly redundant with respect to the optic flow axis. Thus if the optic flow axis can be located, coding gain can be obtained in the presence of motion.

A motion-compensated coder works as follows. An *I* picture is sent, but is also locally stored so that it can be compared with the next input picture to find motion vectors for various areas of the picture. The *I* picture is then shifted according to these vectors to cancel inter-picture motion. The resultant *predicted* picture is compared with the actual picture to produce a *prediction error* also called a *residual*. The prediction error is transmitted with the motion vectors. At the receiver the original *I* picture is also held in a memory. It is shifted according to the transmitted motion vectors to create the predicted picture and then the prediction error is added to it to recreate the original. When a picture is encoded in this way MPEG calls it a *P* picture.

1.7.4   Film-originated video compression

Film can be used as the source of video signals if a telecine machine is used. The most common frame rate for film is 24 Hz, whereas the field rates of television are 50 Hz and 60 Hz. This incompatibility is patched over in two different ways. In 50 Hz telecine, the film is simply played slightly too fast so that the frame rate becomes 25 Hz. Then each frame is converted into two television fields giving the correct 50 Hz field rate. In 60 Hz telecine, the film travels at the correct speed, but alternate frames are used to produce two fields then three fields. The technique is known as 3:2 pulldown. In this way two frames produce five fields and so the correct 60 Hz field rate results. The motion portrayal of telecine is not very good as moving objects judder, especially in 60 Hz systems. Figure 1.10 shows how the optic flow is portrayed in film-originated video.

When film-originated video is input to a compression system, the disturbed optic flow will play havoc with the motion compensation system. In a 50 Hz system there appears to be no motion between the two fields which have originated from the same film frame, whereas between the next two fields large motions will exist. In 60 Hz systems, the motion will be zero for three fields out of five.

With such inputs, it is more efficient to adopt a different processing mode which is based upon the characteristics of the original film. Instead of attempting to manipulate fields of video, the system de-interlaces pairs of fields in order to reconstruct the original film frames. This can be done by a fairly simple motion detector. When substantial motion is measured

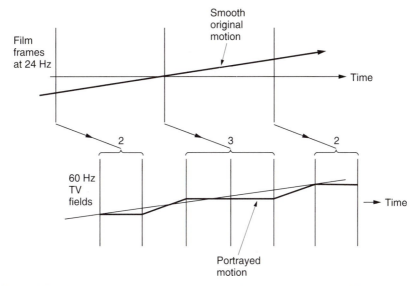

Figure 1.10 Telecine machines must use 3:2 pulldown to produce 60 Hz field rate video.

between successive fields in the output of a telecine, this is taken to mean that the fields have come from different film frames. When negligible motion is detected between fields, this is taken to indicate that the fields have come from the same film frame.

In 50 Hz video it is quite simple to find the sequence and produce de-interlaced frames at 25 Hz. In 60 Hz 3:2 pulldown video the problem is slightly more complex because it is necessary to locate the frames in which three fields are output so that the third field can be discarded, leaving, once more, de-interlaced frames at 25 Hz. While it is relatively straightforward to lock-on to the 3:2 sequence with direct telecine output signals, if the telecine material has been edited on videotape the 3:2 sequence may contain discontinuities. In this case it is necessary to provide a number of field stores in the de-interlace unit so that a series of fields can be examined to locate the edits. Once telecine video has been de-interlaced back to frames, intra- and inter-coded compression can be employed using frame-based motion compensation.

MPEG transmissions include flags which tell the decoder the origin of the material. Material originating at 24 Hz but converted to interlaced video does not have the motion attributes of interlace because the lines in two fields have come from the same point on the time axis. Two fields can be combined to create a progressively scanned frame. In the case of 3:2 pulldown material, the third field need not be sent at all as the decoder can easily repeat a field from memory. As a result the same compressed film material can be output at 50 or 60 Hz as required.

1.8   MPEG-2 profiles and levels

MPEG-2 has too many applications to solve with a single standard and so it is subdivided into Profiles and Levels. Put simply, a Profile describes a degree of complexity whereas a Level describes the picture size or resolution which goes with that Profile. Not all Levels are supported at all Profiles. Figure 1.11 shows the available combinations. In principle there are twenty-four of these, but not all have been defined. An MPEG decoder having a given Profile and Level must also be able to decode lower profiles and levels.

Profiles

		Simple	Main	4:2:2	SNR	Spatial	High
Levels	High		4:2:0 1920 × 1152 90 Mb/S				4:2:0 or 4:2:2 1920 × 1152 100 Mb/S
	High 1440		4:2:0 1440 × 1152 60 Mb/S			4:2:0 1440 × 1152 60 Mb/S	4:2:0 or 4:2:2 1440 × 1152 80 Mb/S
	Main	4:2:0 720 × 576 15 Mb/S NO B	4:2:0 720 × 576 15 Mb/S	4:2:2 720 × 608 50 Mb/S	4:2:0 720 × 576 15 Mb/S		4:2:0 or 4:2:2 720 × 576 20 Mb/S
	Low		4:2:0 352 × 288 4 Mb/S		4:2:0 352 × 288 4 Mb/S		

Figure 1.11   Profiles and levels in MPEG-2. See text for details.

The simple profile does not support bidirectional coding and so only *I* and *P* pictures will be output. This reduces the coding and decoding delay and allows simpler hardware. The simple profile has only been defined at Main level (SP@ML).

The Main Profile is designed for a large proportion of uses. The low level uses a low-resolution input having only 352 pixels per line. The majority of broadcast applications will require the MP@ML (Main Profile at Main Level) subset of MPEG which supports SDTV (standard definition television).

The High-1440 level is a high-definition scheme which doubles the definition compared to main level. The high level not only doubles the resolution but maintains that resolution with 16:9 format by increasing the number of horizontal samples from 1440 to 1920.

In compression systems using spatial transforms and requantizing it is possible to produce scaleable signals. A scaleable process is one in which

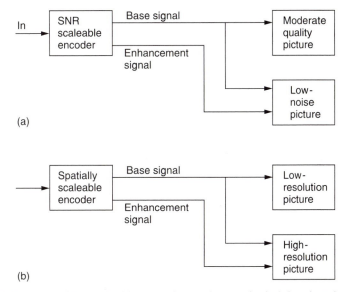

Figure 1.12 (a) An SNR scaleable encoder produces a 'noisy' signal and a noise cancelling signal. (b) A spatally scaleable encoder produces a low-resolution picture and a resolution enhancing picture.

the input results in a main signal and a 'helper' signal. The main signal can be decoded alone to give a picture of a certain quality, but if the information from the helper signal is added some aspect of the quality can be improved.

Figure 1.12(a) shows that in a conventional MPEG coder, by heavily requantizing coefficients a picture with moderate signal-to-noise ratio results. If, however, that picture is locally decoded and subtracted pixel by pixel from the original, a 'quantizing noise' picture would result. This can be compressed and transmitted as the helper signal. A simple decoder only decodes the main 'noisy' bitstream, but a more complex decoder can decode both bitstreams and combine them to produce a low-noise picture. This is the principle of SNR scaleability.

As an alternative, Figure 1.12(b) shows that by coding only the lower spatial frequencies in a HDTV picture a main bitstream can be made which an SDTV receiver can decode. If the lower-definition picture is locally decoded and subtracted from the original picture, a 'definition-enhancing' picture would result. This can be coded into a helper signal. A suitable decoder could combine the main and helper signals to recreate the HDTV picture. This is the principle of spatial scaleability.

The High profile supports both SNR and spatial scaleability as well as allowing the option of 4:2:2 sampling (see section 2.11).

The 4:2:2 profile has been developed for improved compatibility with existing digital television production equipment. This allows 4:2:2

working without requiring the additional complexity of using the high profile. For example, a HP@ML decoder must support SNR scaleability which is not a requirement for production.

1.9 MPEG-2 bitstreams

MPEG-2 supports a variety of bitstream types for various purposes and these are shown in Figure 1.13. The output of a single compressor (video or audio) is known as an Elementary Stream. In transmission, many Elementary Streams will be combined to make a transport stream. A transport stream has a complex structure because it needs to incorporate metadata indicating which audio Elementary Streams and ancillary data are associated with which video Elementary Stream. It is possible to have a single program transport stream (SPTS) which carries only the Elementary Streams of one TV program.

For certain purposes, such as video disc recording, the complexity of a transport stream is not warranted. In this case a program stream can be used. A program stream is a simplified bitstream which multiplexes audio and video for a single program together, provided they have been

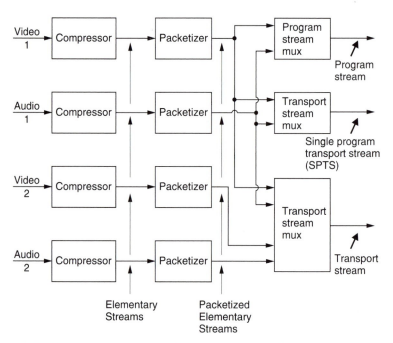

Figure 1.13 The bitstream types of MPEG-2. See text for details.

encoded from a common locked clock. When a recording is played back, the clock can be created by a crystal oscillator and the replay locked to it.

1.10   Drawbacks of compression

By definition, compression removes redundancy from signals. Redundancy is, however, essential to making data resistant to errors. As a result, it is generally true that compressed data are more sensitive to errors than uncompressed data. Thus transmission systems using compressed data must incorporate more powerful error-correction strategies and avoid compression techniques which are notoriously sensitive. As an example, the Digital Betacam format uses relatively mild compression and yet requires 20% redundancy whereas the D-5 format does not use compression and only requires 17% redundancy even though it has a recording density 30% higher.

Techniques using tables such as the Lempel–Ziv–Welch codes are very sensitive to bit errors as an error in the transmission of a table value results in bit errors every time that table location is accessed. This is known as error propagation. Variable-length techniques such as the Huffman code are also sensitive to bit errors. As there is no fixed symbol size, the only way the decoder can parse a serial bitstream into symbols is to increase the assumed wordlength a bit at a time until a code value is recognized. The next bit must then be the first bit in the next symbol. A single bit in error could cause the length of a code to be wrongly assessed and then all subsequent codes would also be wrongly decoded until synchronization could be re-established. Later variable-length codes sacrifice some compression efficiency in order to offer better resynchronization properties.

In non-real-time systems such as computers an uncorrectable error results in reference to the back-up media. In real-time systems such as audio and video this is impossible and concealment must be used. However, concealment relies on redundancy and compression reduces the degree of redundancy. Media such as hard disks can be verified so that uncorrectable errors are virtually eliminated, but tape is prone to dropouts which will exceed the burst correcting power of the replay system from time to time. For this reason the compression factors used on audio or video tape should be moderate.

As perceptive coders introduce noise, it will be clear that in a concatenated system the second codec could be confused by the noise due to the first. If the codecs are identical then each may well make, or better still be designed to make, the same decisions when they are in tandem. If the codecs are not identical the results could be disappointing.

Signal manipulation between codecs can also result in artefacts which were previously undetectible being revealed because the signal which was masking them is no longer present.

In general, compression should not be used for its own sake, but only where a genuine bandwidth or cost bottleneck exists. Even then the mildest compression possible should be used. While high compression factors are permissible for final delivery of material to the consumer, they are not advisable prior to any post-production stages. For contribution material, lower compression factors are essential and this is sometimes referred to as *mezzanine* level compression.

One practical drawback of compression systems is that they are largely generic in structure and the same hardware can be operated at a variety of compression factors. Clearly the higher the compression factor, the cheaper the system will be to operate so there will be economic pressure to use high compression factors. Naturally the risk of artefacts is increased and so there is counterpressure from those with engineering skills to moderate the compression. The way of the world at the time of writing is that the accountants have the upper hand. This was not a problem when there were fixed standards such as PAL and NTSC, as there was no alternative but to adhere to them. Today there is a danger that the variable compression factor control will be turned too far in the direction of economy.

It has been seen above that concatenation of compression systems should be avoided as this causes generation loss. Generation loss is worse if the codecs are different. Interlace is a legacy compression technique and if concatenated with MPEG, generation loss will be exaggerated. In theory and in practice better results are obtained in MPEG for the same bit rate if the input is progressive. Consequently the use of interlace with MPEG coders cannot be recommended for new systems. Chapter 5 explores this theme in greater detail.

1.11   Compression pre-processing

Compression relies completely on identifying redundancy in the source material. Consequently anything which reduces that redundancy will have a damaging effect. Noise is particularly undesirable as it creates additional spatial frequencies in individual pictures as well as spurious differences between pictures. Where noisy source material is anticipated some form of noise reduction will be essential.

When high compression factors must be used to achieve a low bit rate, it is inevitable that the level of artifacts will rise. In order to contain the artifact level, it is necessary to restrict the source entropy prior to the coder. This may be done by spatial low-pass filtering to reduce the picture resolution, and may be combined with downsamplng to reduce the number of pixels per picture. In some cases, such as videophones and teleconferencing it will also be necessary to reduce the picture rate.

A compression pre-processor will combine various types of noise reduction (see Chapter 3) with spatial and temporal downsampling.

1.12　Some guidelines

Although compression techniques themselves are complex, there are some simple rules which can be used to avoid disappointment. Used wisely, MPEG-2 compression has a number of advantages. Used in an inappropriate manner, disappointment is almost inevitable and the technology could get a bad name. The next few points are worth remembering.

- Compression technology may be exciting, but if it is not necessary it should not be used.
- If compression is to be used, the degree of compression should be as small as possible; i.e. use the highest practical bit rate.
- Cascaded compression systems cause loss of quality and the lower the bit rates, the worse this gets. Quality loss increases if any post-production steps are performed between compressions.
- Avoid using interlaced video with MPEG.
- Compression systems cause delay.
- Compression systems work best with clean source material. Noisy signals or poorly decoded composite video give poor results.
- Compressed data are generally more prone to transmission errors than non-compressed data. The choice of a compression scheme must consider the error characteristics of the channel.
- Audio codecs need to be level calibrated so that when sound pressure level-dependent decisions are made in the coder those levels actually exist at the microphone.
- Low bit rate coders should only be used for the final delivery of post-produced signals to the end user.
- Compression quality can only be assessed subjectively.
- Don't believe statements comparing codec performance to 'VHS quality' or similar. Compression artefacts are quite different to the artefacts of consumer VCRs.
- Quality varies wildly with source material. Beware of 'convincing' demonstrations which may use selected material to achieve low bit rates. Use your own test material, selected for a balance of difficulty.
- Don't be browbeaten by the technology. You don't have to understand it to assess the results. Your eyes and ears are as good as

anyone's so don't be afraid to criticize artifacts. In the case of video, use still frames to distinguish spatial artifacts from temporal artifacts.

References

1. ISO/IEC JTC1/SC29/WG11 MPEG, International standard ISO 11172, Coding of moving pictures and associated audio for digital storage media up to 1.5 Mbits/s (1992)
2. Le Gall, D., MPEG: a video compression standard for multimedia applications. *Communications of the ACM*, **34**, No.4, 46–58 (1991)
3. MPEG Video Standard: ISO/IEC 13818–2: Information technology – generic coding of moving pictures and associated audio information: Video (1996) (aka ITU-T Rec. H-262 (1996))
4. Huffman, D.A., A method for the construction of minimum redundancy codes. *Proc. IRE.* **40** 1098–1101 (1952)
5. ISO Joint Photographic Experts Group standard JPEG-8-R8
6. Wallace, G.K., Overview of the JPEG (ISO/CCITT) still image compression standard. ISO/JTC1/SC2/WG8 N932 (1989)

2

Fundamentals

2.1 What is an audio signal?

Actual sounds are converted to electrical signals for convenience of handling, recording and conveying from one place to another. This is the job of the microphone. There are two basic types of microphone, those which measure the variations in air pressure due to sound, and those which measure the air velocity due to sound, although there are numerous practical types which are a combination of both.

The sound pressure or velocity varies with time and so does the output voltage of the microphone, in proportion. The output voltage of the microphone is thus an analog of the sound pressure or velocity.

As sound causes no overall air movement, the average velocity of all sounds is zero, which corresponds to silence. As a result the bidirectional air movement gives rise to bipolar signals from the microphone, where silence is in the centre of the voltage range, and instantaneously negative or positive voltages are possible. Clearly the average voltage of all audio signals is also zero, and so when level is measured, it is necessary to take the modulus of the voltage, which is the job of the rectifier in the level meter. When this is done, the greater the amplitude of the audio signal, the greater the modulus becomes, and so a higher level is displayed.

While the nature of an audio signal is very simple, there are many applications of audio, each requiring different bandwidth and dynamic range.

2.2 What is a video signal?

The goal of television is to allow a moving picture to be seen at a remote place. The picture is a two-dimensional image, which changes as a function of time. This is a three-dimensional information source where the dimensions are distance across the screen, distance down the screen and time.

While telescopes convey these three dimensions directly, this cannot be done with electrical signals or radio transmissions, which are restricted to a single parameter varying with time. The solution in film and television is to convert the three-dimensional moving image into a series of still pictures, taken at the frame rate, and then, in television only, the two-dimensional images are scanned as a series of lines to produce a single voltage varying with time which can be recorded or transmitted. Europe, the Middle East and the former Soviet Union use the scanning standard of 625/50, whereas the USA and Japan use 525/59.94.

2.3   Types of video

Figure 2.1 shows some of the basic types of analog colour video. Each of these types can, of course, exist in a variety of line standards. Since practical colour cameras generally have three separate sensors, one for each primary colour, an *RGB* component system will exist at some stage in the internal workings of the camera, even if it does not emerge in that form. *RGB* consists of three parallel signals each having the same spectrum, and is used where the highest accuracy is needed, often for production of still pictures. Examples of this are paint systems and in computer-aided design (CAD) displays. *RGB* is seldom used for real-time video recording.

Some compression can be obtained by using colour difference working. The human eye relies on brightness to convey detail, and much less resolution is needed in the colour information. *R*, *G* and *B* are matrixed together to form a luminance (and monochrome compatible) signal *Y* which has full bandwidth. The matrix also produces two colour difference signals, *R-Y* and *B-Y*, but these do not need the same bandwidth as *Y*, one half or one quarter will do depending on the application. Colour difference signals represent an early application of perceptive coding; a saving in bandwidth is obtained by expressing the signals according to the way the eye operates. Analog colour difference recorders such as Betacam and M II record these signals separately. The D-1 and D-5 formats records 525/60 or 625/50 colour difference signals digitally and Digital Betacam does so using compression. In casual parlance, colour difference formats are often called component formats to distinguish them from composite formats.

For colour television broadcast in a single channel, the PAL, SECAM and NTSC systems interleave into the spectrum of a monochrome signal a subcarrier which carries two colour difference signals of restricted

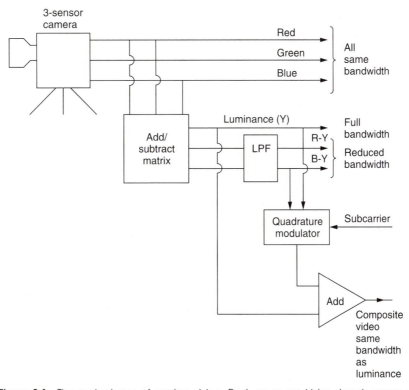

Figure 2.1 The major types of analog video. Red, green and blue signals emerge from the camera sensors, needing full bandwidth. If a luminance signal is obtained by a weighted sum of *R*, *G* and *B*, it will need full bandwidth, but the colour difference signals *R–Y* and *B–Y* need less bandwidth. Combining *R–Y* and *B–Y* into a subcarrier modulation scheme allows colour transmission in the same bandwidth as monochrome.

bandwidth. As the bandwidth required for composite video is no greater than that of luminance, it can be regarded as a form of compression performed in the analog domain. The artifacts which composite video introduces and the inflexibility in editing resulting from the need to respect colour framing serve as a warning that compression is not without its penalties. The subcarrier is intended to be invisible on the screen of a monochrome television set. A subcarrier-based colour system is generally referred to as composite video, and the modulated subcarrier is called chroma.

It is not advantageous to compress composite video using modern transform-based coders as the transform process cannot identify redundancy in a subcarrier. Composite video compression is restricted to differential coding systems. Transform-based compression must use *RGB* or colour difference signals. As *RGB* requires excessive bandwidth it makes no sense to use it with compression and so in practice only colour difference signals, which have been bandwidth reduced by perceptive coding, are used in MPEG-2. Where signals to be compressed originate in composite form, they must be decoded first. The decoding must be performed as accurately as possible, with particular attention being given to the quality of the *Y/C* separation. The chroma in composite signals is deliberately designed to invert from frame to frame in order to lessen its visibility. Unfortunately any residual chroma in luminance will be interpreted by inter-field compression systems as temporal luminance changes which need to be reproduced. This eats up data which should be used to render the picture. Residual chroma also results in high horizontal and vertical spatial frequencies in each field which appear to be wanted detail to the compressor.

2.4   What is a digital signal?

One of the vital concepts to grasp is that digital audio and video are simply alternative means of carrying the same information as their analog counterparts. An ideal digital system has the same characteristics as an ideal analog system: both of them are totally transparent and reproduce the original applied waveform without error. Needless to say, in the real world ideal conditions seldom prevail, so analog and digital equipment both fall short of the ideal. Digital equipment simply falls short of the ideal to a smaller extent than does analog and at lower cost, or, if the designer chooses, can have the same performance as analog at much lower cost. Compression is one of the techniques used to lower the cost, but it has the potential to lower the quality as well.

Any analog signal source can be characterized by a given useful bandwidth and signal-to-noise ratio. Video signals have very wide bandwidth extending over several megahertz but require only 50 dB or so SNR whereas audio signals require only 20 kHz but need much better SNR.

Although there are a number of ways in which audio and video waveforms can be represented digitally, there is one system, known as Pulse Code Modulation (PCM), which is in virtually universal use. Figure 2.2 shows how PCM works. Instead of being continuous, the time axis is represented in a discrete, or stepwise manner. The waveform is not carried by continuous representation, but by measurement at regular intervals. This process is called sampling and the frequency with which samples are taken is called the sampling rate or sampling frequency F_s. The sampling rate is generally fixed and is not necessarily a function of any frequency in the signal, although in component video it will be line-locked for convenience. If every effort is made to rid the sampling clock of jitter, or time instability, every sample will be made at an exactly even time step. Clearly if there is any subsequent timebase error, the instants at which samples arrive will be changed and the effect can be detected. If samples arrive at some destination with an irregular timebase, the effect can be eliminated by storing the samples temporarily in a memory and reading them out using a stable, locally generated clock. This process is called timebase correction which all properly engineered digital systems employ. It should be stressed that sampling is an analog process. Each sample still varies infinitely as the original waveform did.

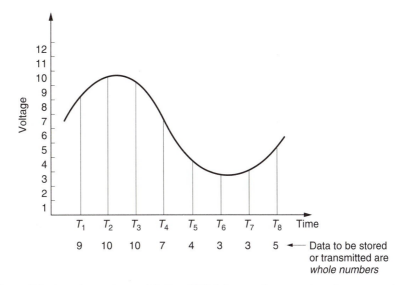

Figure 2.2 In pulse code modulation (PCM) the analog waveform is measured periodically at the sampling rate. The voltage (represented here by the height) of each sample is then described by a whole number. The whole numbers are stored or transmitted rather than the waveform itself.

Those who are not familiar with the technology often worry that sampling takes away something from a signal because it is not taking notice of what happened between the samples. This would be true in a system having infinite bandwidth, but no analog signal can have infinite bandwidth. All analog signal sources from microphones, tape decks, cameras and so on have a frequency response limit, as indeed do our ears and eyes. When a signal has finite bandwidth, the rate at which it can change is limited, and the way in which it changes becomes predictable. When a waveform can only change between samples in one way, it is then only necessary to carry the samples and the original waveform can be reconstructed from them.

Figure 2.2 also shows that each sample is also discrete, or represented in a stepwise manner. The length of the sample, which will be proportional to the voltage of the waveform, is represented by a whole number. This process is known as quantizing and results in an approximation, but the size of the error can be controlled until it is negligible. If, for example we were to measure the height of humans to the nearest metre, virtually all adults would register two metres high and obvious difficulties would result. These are generally overcome by measuring height to the nearest centimetre. Clearly there is no advantage in going further and expressing our height in a whole number of millimetres or even micrometres. An appropriate resolution can be found just as readily for audio or video, and greater accuracy is not beneficial. The link between quality and sample resolution is explored later in this chapter. The advantage of using whole numbers is that they are not prone to drift. If a whole number can be carried from one place to another without numerical error, it has not changed at all. By describing waveforms numerically, the original information has been expressed in a way which is better able to resist unwanted changes.

Essentially, digital systems carry the original waveform numerically. The number of the sample is an analog of time, and the magnitude of the sample is an analog of the signal voltage. As both axes of the waveform are discrete, the waveform can be accurately restored from numbers as if it were being drawn on graph paper. If we require greater accuracy, we simply choose paper with smaller squares. Clearly more numbers are required and each one could change over a larger range.

Discrete numbers are used to represent the value of samples so that they can readily be transmitted or processed by binary logic. There are two ways in which binary signals can be used to carry sample data. When each digit of the binary number is carried on a separate wire this is called parallel transmission. The state of the wires changes at the sampling rate. This approach is used in the parallel video interfaces, as video needs a relatively short wordlength; eight or ten bits. Using multiple wires is cumbersome where a long wordlength is in use, and a single wire can be used where successive digits from each sample are sent serially. This is the definition of Pulse Code Modulation. Clearly the clock frequency must now be higher than the sampling rate.

If a well-engineered PCM digital channel having a wider bandwidth and a greater signal-to-noise ratio is put in series with an analog source, it is only necessary to set the levels correctly and the analog signal is then subject to no loss of information whatsoever. The digital clipping level is above the largest analog signal, the digital noise floor is below the inherent noise in the signal and the low- and high-frequency response of the digital channel extends beyond the frequencies in the analog signal.

Digital signals of this form will be used as the input to compression systems and must also be output by the decoding stage in order that the signal can be returned to analog form. Figure 2.3 shows the stages involved. Between the coder and the decoder the signal is not PCM but

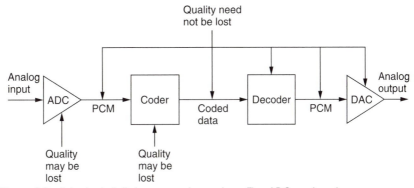

Figure 2.3 A typical digital compression system. The ADC and coder are responsible for most of the quality loss, whereas the PCM and coded data channels cause no further loss (excepting bit errors).

will be in a format which is highly dependent on the kind of compression technique used. It will also be evident from Figure 2.3 where the signal quality of the system can be impaired. The PCM digital interfaces between the ADC and the coder and between the decoder and the DAC cause no loss of quality. Quality is determined by the ADC and by the performance of the coder. Generally decoders do not cause significant loss of quality; they make the best of the data from the coder. Similarly, DACs cause little quality loss above that due to the ADC. In practical systems the loss of quality is dominated by the action of the coder. In communication theory, compression is known as *source coding* in order to distinguish it from the *channel coding* necessary reliably to send data down transmission or recording channels. This book is not concerned with channel coding but details can be found elsewhere.[1]

2.5   Introduction to conversion

There are a number of ways in which a waveform can be digitally represented, but the most useful and therefore common is PCM as described above. The input is a continuous-time, continuous-voltage waveform, and this is converted into a discrete-time, discrete-voltage format by a combination of sampling and quantizing. As these two processes are orthogonal (a $64 word meaning at right angles to one another) they are totally independent and can be performed in either order. Figure 2.4(a) shows an analog sampler preceding a quantizer, whereas (b) shows an asynchronous quantizer preceding a digital sampler. Ideally, both will give the same results; in practice each suffers from different deficiencies. Both approaches will be found in real equipment.

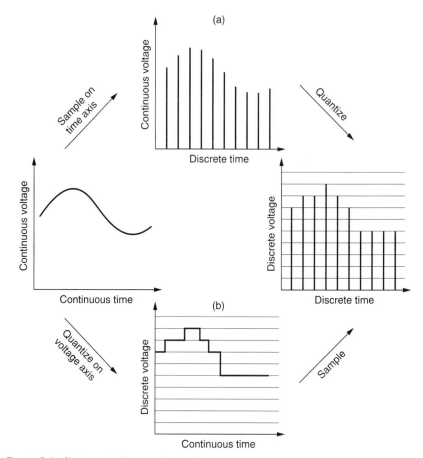

Figure 2.4 Since sampling and quantizing are orthogonal, the order in which they are performed is not important. In (a) sampling is performed first and the samples are quantized. This is common in audio converters. In (b) the analog input is quantized into an asynchronous binary code. Sampling takes place when this code is latched on sampling clock edges. This approach is universal in video converters.

The independence of sampling and quantizing allows each to be discussed quite separately in some detail, prior to combining the processes for a full understanding of conversion.

2.6   Sampling and aliasing

Sampling is no more than periodic measurement, and it will be shown here that there is no theoretical need for sampling to be audible or visible. Practical equipment may, of course be less than ideal, but, given good engineering practice, the ideal may be approached quite closely.

Sampling must be precisely regular, because the subsequent process of timebase correction assumes a regular original process. The sampling process originates with a pulse train which is shown in Figure 2.5(a) to be of constant amplitude and period. The signal waveform amplitude-modulates the pulse train in much the same way as the carrier is modulated in an AM radio transmitter. One must be careful to avoid over-modulating the pulse train as shown in (b) and in the case of bipolar signals this is helped by applying a DC offset to the analog waveform so that silence corresponds to a level half-way up the pulses as at (c). Clipping due to any excessive input level will then be symmetrical.

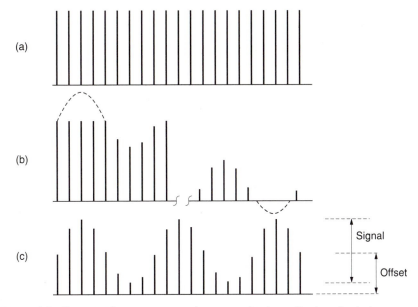

Figure 2.5 The sampling process requires a constant-amplitude pulse train as shown in (a). This is amplitude modulated by the waveform to be sampled. If the input waveform has excessive amplitude or incorrect level, the pulse train clips as shown in (b). For a bipolar waveform, the greatest signal level is possible when an offset of half the pulse amplitude is used to centre the waveform as shown in (c).

In the same way that AM radio produces sidebands or images above and below the carrier, sampling also produces sidebands although the carrier is now a pulse train and has an infinite series of harmonics as can be seen in Figure 2.6(a). The sidebands in (b) repeat above and below each harmonic of the sampling rate.

The sampled signal can be returned to the continuous-time domain simply by passing it into a low-pass filter. This filter has a frequency response which prevents the images from passing, and only the baseband signal emerges, completely unchanged.

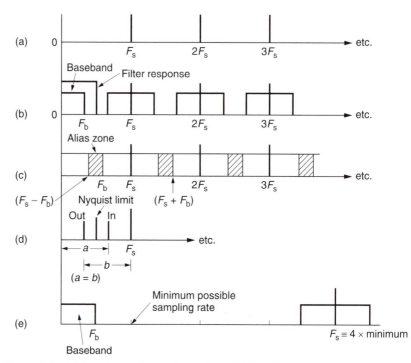

Figure 2.6 (a) Spectrum of sampling pulses. (b) Spectrum of samples. (c) Aliasing due to sideband overlap. (d) Beat-frequency production. (e) 4× oversampling.

If an input is supplied having an excessive bandwidth for the sampling rate in use, the sidebands will overlap, and the result is aliasing, where certain output frequencies are not the same as their input frequencies but become difference frequencies. It will be seen from Figure 2.6(c) that aliasing occurs when the input frequency exceeds half the sampling rate, and this derives the most fundamental rule of sampling, which is that the sampling rate must be at least twice the input bandwidth.

In addition to the low-pass filter needed at the output to return to the continuous-time domain, a further low-pass filter is needed at the input to prevent aliasing. If input frequencies of more than half the sampling rate cannot reach the sampler, aliasing cannot occur.

While aliasing has been described above in the frequency domain, it can equally be described in the time domain. In Figure 2.7(a) the sampling rate is obviously adequate to describe the waveform, but at (b) it is inadequate and aliasing has occurred.

Aliasing is commonly seen on television and in the cinema, owing to the relatively low frame rates used. With a frame rate of 24 Hz, a film camera will alias on any object changing at more than 12 Hz. Such objects include the spokes of stagecoach wheels, aircraft propellors and helicopter rotors. When the spoke- or blade-passing frequency reaches 24 Hz the motion appears to stop.

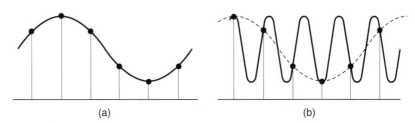

Figure 2.7 In (a), the sampling is adequate to reconstruct the original signal. In (b) the sampling rate is inadequate and reconstruction produces the wrong waveform (detailed). Aliasing has taken place.

In television systems the input image which falls on the camera sensor will be continuous in time, and continous in two spatial dimensions corresponding to the height and width of the sensor. All three of these continuous dimensions will be sampled. There is a direct connection between the concept of temporal sampling, where the input signal changes with respect to time at some frequency and is sampled at some other frequency, and spatial sampling, where an image changes a given number of times per unit distance and is sampled at some other number of times per unit distance. The connection between the two is the process of scanning. Temporal frequency can be obtained by multiplying spatial frequency by the speed of the scan. Figure 2.8 shows a hypothetical image sensor which has one thousand discrete sensors across a width of one centimetre. The spatial sampling rate of this sensor is thus one thousand per centimetre. If the sensors are measured sequentially during a scan which takes one millisecond to go across the one centimetre width, the result will be a temporal sampling rate of 1 MHz.

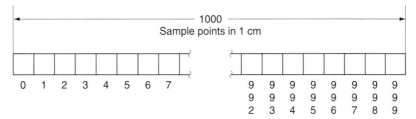

Figure 2.8 If the above spatial sampling arrangement of 1000 points per centimetre is scanned in 1 millisecond, the sampling rate will become 1 megahertz.

2.7   Reconstruction

If ideal low-pass anti-aliasing and anti-image filters are assumed, having a vertical cut-off slope at half the sampling rate, an ideal spectrum shown at Figure 2.9(a) is obtained. Figure 2.9(b) shows that the impulse response of a phase linear ideal low-pass filter is a sin$x$/x waveform in the time domain. Such a waveform passes through zero volts periodically. If the cut-off frequency of the filter is one-half of the sampling rate, the impulse passes through zero *at the sites of all other samples*. Thus at the output of such a filter, the voltage at the centre of a sample is due to that sample alone, since the value of *all* other samples is zero at that instant. In other words the continuous time output waveform must join up the tops of the

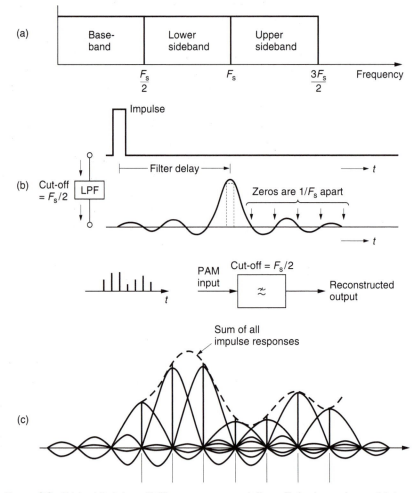

Figure 2.9 If ideal 'brick wall' filters are assumed, the efficient spectrum of (a) results. An ideal low-pass filter has an impulse response shown in (b). The impulse passes through zero at intervals equal to the sampling period. When convolved with a pulse train at the sampling rate, as shown in (c), the voltage at each sample instant is due to that sample alone as the impulses from all other samples pass through zero there.

input samples. In between the sample instants, the output of the filter is the sum of the contributions from many impulses, and the waveform smoothly joins the tops of the samples. If the time domain is being considered, the anti-image filter of the frequency domain can equally well be called the reconstruction filter. It is a consequence of the band-limiting of the original anti-aliasing filter that the filtered analog waveform could only travel between the sample points in one way. As the reconstruction filter has the same frequency response, the reconstructed output waveform must be identical to the original band-limited waveform prior to sampling. A rigorous mathematical proof of the above has been available since the 1930s, when PCM was invented, and can also be found in Betts.[2]

2.8   Filter design

The ideal filter with a vertical 'brick-wall' cut-off slope is difficult to implement. As the slope tends to vertical, the delay caused by the filter goes to infinity: the quality is marvellous but you don't live to measure it. In practice, a filter with a finite slope is accepted as shown in Figure 2.10, and the sampling rate has to be raised a little to prevent aliasing. There is no absolute factor by which the sampling rate must be raised; it depends upon the filters which are available.

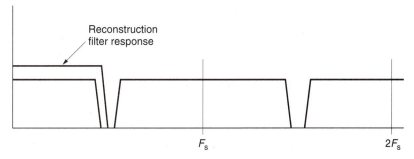

Figure 2.10   As filters with finite slope are needed in practical systems, the sampling rate is raised slightly beyond twice the highest frequency in the basband.

It is not easy to specify such filters, particularly the amount of stopband rejection needed. The amount of aliasing resulting would depend on, among other things, the amount of out-of-band energy in the input signal. This is seldom a problem in video, but can warrant attention in audio where overspecified bandwidths are sometimes found. As a further complication, an out-of-band signal will be attenuated by the response of the anti-aliasing filter to that frequency, but the residual signal will then alias, and the reconstruction filter will reject it according to its attenuation at the new frequency to which it has aliased.

It could be argued that the reconstruction filter is unnecessary in audio, since all the images are outside the range of human hearing. However, the slightest non-linearity in subsequent stages would result in gross intermodulation distortion. The possibility of damage to tweeters and beating with the bias systems of analog tape recorders must also be considered. In video the filters are essential to constrain the bandwidth of the signal to the allowable broadcast channel width.

Figure 2.11 shows the terminology used to describe the common elliptic low-pass filter. These filters are popular because they can be realized with fewer components than other filters of similar response. It is a characteristic of these elliptic filters that there are ripples in the passband and stopband. In much equipment the anti-aliasing filter and the reconstruction filter will have the same specification, so that the passband ripple is doubled with a corresponding increase in dispersion. Sometimes slightly different filters are used to reduce the effect.

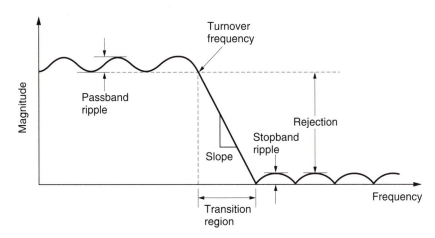

Figure 2.11 The important features and terminology of low-pass filters used for anti-aliasing and reconstruction.

It is difficult to produce an analog filter with low distortion. Passive filters using inductors suffer non-linearity at high levels due to the B/H curve of the cores. Active filters can simulate inductors which are linear using op-amp techniques, but they tend to suffer non-linearity at high frequencies where the falling open-loop gain reduces the effect of feedback.

Since a sharp cut-off is generally achieved by cascading many filter sections which cut at a similar frequency, the phase responses of these sections will accumulate. The phase may start to leave linearity at a fraction of the cut-off frequency and by the time this is reached, the phase may have completed several revolutions. Meyer[3] suggests that these phase errors are audible and that equalization is necessary. In video, phase linearity is essential as otherwise the different frequency components of a contrast step are smeared across the screen. An advantage of linear phase filters is that ringing is minimized, and there is less possibility of clipping on transients.

It is possible with substantial expense and effort to construct a ripple-free phase-linear filter with the required stopband rejection,[4,5] but the requirement can be eliminated by the use of oversampling. As shown in Figure 2.12 a high sampling rate produces a large spectral gap between the baseband and the first lower sideband. The anti-aliasing and reconstruction filters need only have a gentle roll-off, causing minimum disturbance to phase linearity in the baseband, and the Butterworth configuration, which does not have ripple or dispersion, can be used. The penalty of oversampling is that an excessive data rate results. It is necessary to reduce the rate following the ADC using a digital LPF. The sampling rate can subsequently be raised prior

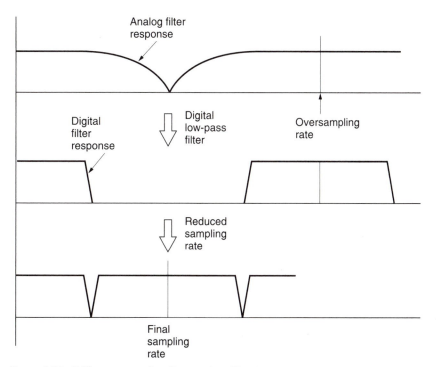

Figure 2.12 With oversampling the analog filters become non-critical since the passband is determined by digital filters.

to the DAC using an interpolator. Digital filters can be made perfectly phase linear and, using LSI, can be inexpensive to construct. The superiority of oversampling converters means that they have become universal in audio and are starting to do so in video.

2.9   Sampling clock jitter

The instants at which samples are taken in an ADC and the instants at which DACs make conversions must be evenly spaced, otherwise unwanted signals can be added to the waveform. Figure 2.13(a) shows the effect of sampling clock jitter on a sloping waveform. Samples are taken at the wrong times. When these samples have passed through a system, the timebase correction stage prior to the DAC will remove the jitter, and the result is shown in (b). The magnitude of the unwanted signal is proportional to the slope of the signal waveform and so increases with frequency. The nature of the unwanted signal depends on the spectrum of the jitter. If the jitter is random, the effect is noise-like and relatively benign unless the amplitude is excessive. Clock jitter is, however, not necessarily random.

Figure 2.14 shows that one source of clock jitter is crosstalk on the clock signal. The unwanted additional signal changes the time at which the sloping clock signal appears to cross the threshold voltage of the clock receiver. The threshold itself may be changed by ripple on the clock receiver power supply.

In MPEG systems the decoder clock is derived from the incoming transport stream which may be subject to jitter induced by the transmission path. If the jitter filtering is inadequate the entire timebase of the decoder may be unstable.

The allowable jitter is measured in picoseconds in both audio and video signals, as shown in Figure 2.13, and clearly steps must be taken to eliminate it by design. Convertor clocks must be generated from clean power supplies which are well decoupled from the power used by the logic because a convertor clock must have a good signal to noise ratio. If an external clock is used, it cannot be used directly, but must be fed through a well-damped phase-locked-loop which will filter out the jitter. The external clock signal is sometimes fed into the clean circuitry using an optical coupler to improve isolation.

Although it has been documented for many years, attention to control of clock jitter is not as great in actual audio and video hardware as it might be. It accounts for many of the slight differences between convertors reproducing the same data. A well-engineered convertor should substantially reject jitter on an external clock and should sound

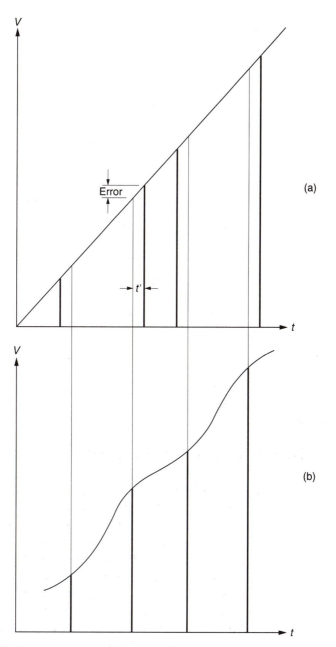

Figure 2.13 The effect of sampling timing jitter on noise, and calculation of the required accuracy for a 16-bit system. (a) Ramp sampled with jitter has error proportional to slope. (b) When jitter is removed by later circuits, error appears as noise added to samples. For a 16-bit system there are $2^{16}Q$, and the maximum slope at 20 KHz will be $20\,000\pi \times 2^{16}Q$ per second. If jitter is to be neglected, the noise must be less than $\tfrac{1}{2}Q$, thus timing accuracy t' multiplied by maximum slope $= \tfrac{1}{2}Q$ or $20\,000\pi \times 2^{16}Qt' = \tfrac{1}{2}Q$

$$\therefore \; t' = \frac{1}{2 \times 20\,000 \times \pi \times 2^{16}} = 121\,\text{ps}$$

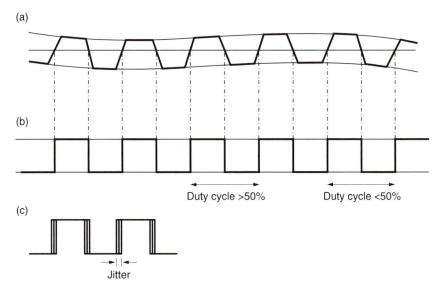

(a)

(b)

Duty cycle >50% Duty cycle <50%

(c)

Jitter

Figure 2.14 Crosstalk in transmission can result in unwanted signals being added to the clock waveform. It can be seen here that a low-frequency interference signal affects the slicing of the clock and causes a periodic jitter.

the same when reproducing the same data irrespective of the source of the data.

2.10   Choice of audio sampling rate

The Nyquist criterion is only the beginning of the process which must be followed to arrive at a suitable sampling rate. The slope of available filters will compel designers to raise the sampling rate above the theoretical Nyquist rate. For consumer products, the lower the sampling rate, the better, since the cost of the medium is directly proportional to the sampling rate: thus sampling rates near to twice 20 KHz are to be expected. For professional products, there is a need to operate at variable speed for pitch correction. When the speed of a digital recorder is reduced, the offtape sampling rate falls, and Figure 2.15 shows that with a minimal sampling rate the first image frequency can become low enough to pass the reconstruction filter. If the sampling frequency is raised without changing the response of the filters, the speed can be reduced without this problem. It follows that variable-speed recorders, generally those with stationary heads, must use a higher sampling rate.

In the early days of digital audio, video recorders were adapted to store audio samples by creating a pseudo-video waveform which could convey

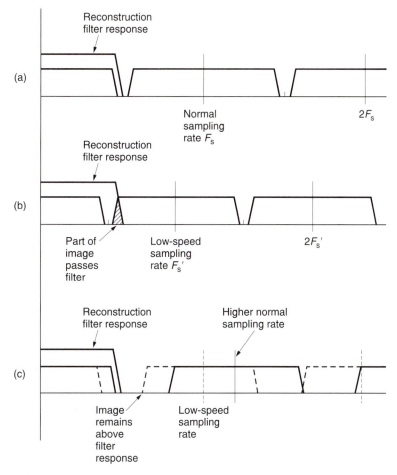

Figure 2.15 At normal speed, the reconstruction filter correctly prevents images entering the baseband, as in (a). When speed is reduced, the sampling rate falls, and a fixed filter will allow part of the lower sideband of the sampling frequency to pass. If the sampling rate of the machine is raised, but the filter characteristic remains the same, the problem can be avoided, as in (c).

binary as black and white levels.[6] The sampling rate of such a system is constrained to relate simply to the field rate and field structure of the television standard used, so that an integer number of samples can be stored on each usable TV line in the field. Such a recording can be made on a monochrome recorder, and these recordings are made in two standards, 525 lines at 60 Hz and 625 lines at 50 Hz. Thus it was necessary to find a frequency which is a common multiple of the two and also suitable for use as a sampling rate.

The allowable sampling rates in a pseudo-video system can be deduced by multiplying the field rate by the number of active lines in a field (blanked lines cannot be used) and again by the number of samples in a line. By careful choice of parameters it is possible to use either 525/60 or 625/50 video with a sampling rate of 44.1 kHz.

In 60 Hz video, there are 35 blanked lines, leaving 490 lines per frame, or 245 lines per field for samples. If three samples are stored per line, the sampling rate becomes

$$60 \times 245 \times 3 = 44.1 \, \text{kHz}$$

In 50 Hz video, there are 37 lines of blanking, leaving 588 active lines per frame, or 294 per field, so the same sampling rate is given by

$$50 \times 294 \times 3 = 44.1 \, \text{kHz}$$

The sampling rate of 44.1 kHz came to be that of the Compact Disc. Even though CD has no video circuitry, the equipment used to make CD masters is video based and determines the sampling rate.

For landlines to FM stereo broadcast transmitters having a 15 kHz audio bandwidth, the sampling rate of 32 kHz is more than adequate, and has been in use for some time in the United Kingdom and Japan. This frequency is also in use in the NICAM 728 stereo TV sound system. The professional sampling rate of 48 kHz was proposed as having a simple relationship to 32 kHz, being far enough above 40 kHz for variable-speed operation, and having a simple relationship with PAL video timing which would allow digital video recorders to store the convenient number of 960 audio samples per video field. This is the sampling rate used by all production DVTRs. The field rate offset of NTSC does not easily relate to any of the above sampling rates, and requires special handling which is outside the scope of this book.[7]

Although in a perfect world the adoption of a single sampling rate might have had virtues, for practical and economic reasons digital audio now has essentially three rates to support: 32 kHz for broadcast, 44.1 kHz for CD, and 48 kHz for professional use.[8] Variations of the DAT format will support all of these rates, although the 32 kHz version is uncommon. In MPEG-2 these audio sampling rates may be halved for low-bit rate applications with a corresponding loss of audio bandwidth.

2.11 Video sampling structures

Component or colour difference signals are used primarily for post-production work where quality and flexibility are paramount. In colour difference working, the important requirement is for image manipulation in the digital domain. This is facilitated by a sampling rate which is a multiple of line rate because then there is a whole number of samples in a line and samples are always in the same position along the line and can form neat columns. A practical difficulty is that the line period of the 525 and 625 systems is slightly different. The problem was overcome by the use of a sampling clock which is an integer multiple of both line rates.

ITU-601 (formerly CCIR-601) recommends the use of certain sampling rates which are based on integer multiples of the carefully chosen fundamental frequency of 3.375 MHz. This frequency is normalized to 1 in the document.

In order to sample 625/50 luminance signals without quality loss, the lowest multiple possible is 4 which represents a sampling rate of 13.5 MHz. This frequency line-locks to give 858 samples per line period in 525/59.94 and 864 samples per line period in 625/50.

In the component analog domain, the colour difference signals used for production purposes typically have one half the bandwidth of the luminance signal. Thus a sampling rate multiple of 2 is used and results in 6.75 MHz. This sampling rate allows respectively 429 and 432 samples per line.

Component video sampled in this way has a 4:2:2 format. While other combinations are possible, 4:2:2 is the format for which the majority of digital component production equipment is constructed and is the only component format for which parallel and serial interface standards exist. The D-1, D-5, D-9, SX and Digital Betacam DVTRs operate with 4:2:2 format data. Figure 2.16 shows the spatial arrangement given by 4:2:2

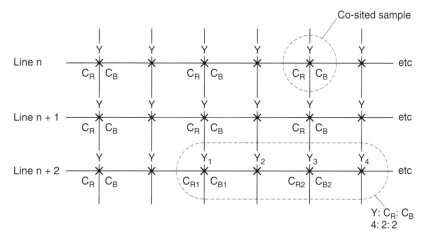

Figure 2.16 In CCIR-601 sampling mode 4:2:2, the line synchronous sampling rate of 13.5 MHz results in samples having the same position in successive lines, so that vertical columns are generated. The sampling rates of the colour difference signals C_R, C_B are one-half of that of luminance, i.e. 6.75 MHz, so that there are alternate Y only samples and co-sited samples which describe Y, C_R and C_B. In a run of four samples, there will be four Y samples, two C_R samples and two C_B samples, hence 4:2:2.

sampling. Luminance samples appear at half the spacing of colour difference samples, and every other luminance sample is co-sited with a pair of colour difference samples. Co-siting is important because it allows all attributes of one picture point to be conveyed with a three-sample vector quantity. Modification of the three samples allows such techniques as colour correction to be performed. This would be difficult without co-sited information. Co-siting is achieved by clocking the three ADCs simultaneously.

For lower bandwidths, particularly in prefiltering operations prior to compression, the sampling rate of the colour difference signal can be halved. 4:1:1 delivers colour bandwidth in excess of that required by the composite formats.

In 4:2:2 the colour difference signals are sampled horizontally at half the luminance sampling rate, yet the vertical colour difference sampling rates are the same as for luminance. While this is not a problem in a production application, this disparity of sampling rates represents a data rate overhead which is undesirable in a compression environment. In this case it is possible to halve the vertical sampling rate of the colour difference signals as well. Figure 2.17 shows that in MPEG-2 4:2:0 sampling, the colour difference signals are downsampled so that the same vertical and horizontal resolution is obtained.

The chroma samples in 4:2:0 are positioned half-way between luminance samples in the vertical axis so that they are evenly spaced when an interlaced source is used. To obtain a 4:2:2 output from 4:2:0 data a vertical interpolation process will be needed in addition to low-pass filtering.

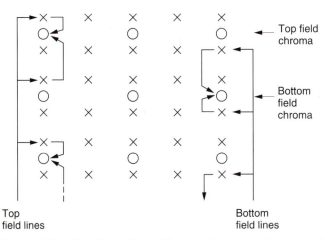

Figure 2.17 In 4:2:0 coding the colour difference pixels are downsampled vertically as well as horizontally. Note that the sample sites need to be vertically interpolated so that when two interlaced fields are combined the spacing is even.

The sampling rates of ITU-601 are based on commonality between 525- and 625-line systems. However, the consequence is that the pixel spacing is different in the horizontal and vertical axes. This is incompatible with computer graphics in which so-called 'square' pixels are used. This means that the horizontal and vertical spacing is the same, giving the same resolution in both axes. However, high-definition TV and computer graphics formats universally use 'square' pixels. MPEG can handle various pixel aspect ratios and allows a control code to be embedded in the sequence header to help the decoder.

2.12   The phase-locked loop

All digital video systems need to be clocked at the appropriate rate in order to function properly. While a clock may be obtained from a fixed frequency oscillator such as a crystal, many operations in video require *genlocking* or synchronizing the clock to an external source.

In phase-locked loops, the oscillator can run at a range of frequencies according to the voltage applied to a control terminal. This is called a voltage-controlled oscillator or VCO. Figure 2.18 shows that the VCO is driven by a phase error measured between the output and some reference. The error changes the control voltage in such a way that the error is reduced, such that the output eventually has the same frequency as the reference. A low-pass filter is fitted in the control voltage path to prevent the loop becoming unstable. If a divider is placed between the VCO and the phase comparator, as in the figure, the VCO frequency can be made to be a multiple of the reference. This also has the effect of making the loop more heavily damped, so that it is less likely to change frequency if the input is irregular.

In digital video, the frequency multiplication of a phase-locked loop is extremely useful. Figure 2.19 shows how the 13.5 MHz clock of

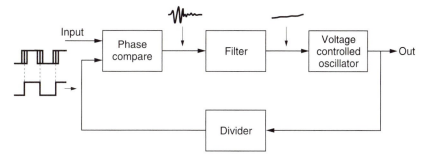

Figure 2.18 A phase-locked loop requires these components as a minimum. The filter in the control voltage serves to reduce clock jitter.

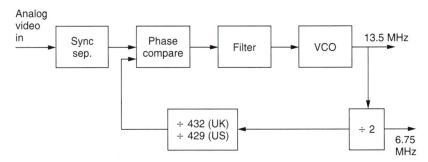

Figure 2.19 In order to obtain 13.5 MHz from input syncs, a PLL with an appropriate division ratio is required.

component digital video is obtained from the sync. pulses of an analog reference by such a multiplication process.

The numerically locked loop is a digital relative of the phase-locked loop. Figure 2.20 shows that the input is an intermittently transmitted value from a counter. The input count is compared with the value of a local count and the difference is used to control the frequency of a local oscillator. Once lock is achieved, the local oscillator and the remote oscillator will run at exactly the same frequency even though there is no continuous link between them.

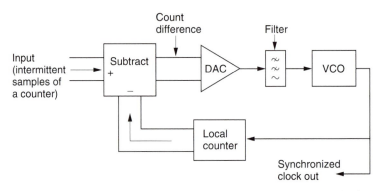

Figure 2.20 In the numerically locked loop, the state of a master counter is intermittently sent. A local oscillator is synchronized to the master oscillator by comparing the states of the local and master counters.

2.13 Quantizing

Quantizing is the process of expressing some infinitely variable quantity by discrete or stepped values. Quantizing turns up in a remarkable

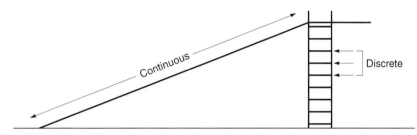

Figure 2.21 An analog parameter is continuous whereas a quantized parameter is restricted to certain values. Here the sloping side of a ramp can be used to obtain any height whereas a ladder only allows discrete heights.

number of everyday guises. Figure 2.21 shows that an inclined ramp enables infinitely variable height to be achieved, whereas a step-ladder allows only discrete heights to be had. A step-ladder quantizes height. When accountants round off sums of money to the nearest pound or dollar they are quantizing.

In audio the values to be quantized are infinitely variable voltages from an analog source. Strict quantizing is a process which is restricted to the voltage domain only. For the purpose of studying the quantizing of a single sample, time is assumed to stand still. This is achieved in practice either by the use of a track-hold circuit or the adoption of a quantizer technology which operates before the sampling stage.

Figure 2.22(a) shows that the process of quantizing divides the voltage range up into quantizing intervals Q. In applications such as telephony these may be of differing size, but for digital audio and video the quantizing intervals are made as identical as possible. If this is done, the binary numbers which result are truly proportional to the original analog voltage, and the digital equivalents of filtering and gain changing can be performed by adding and multiplying sample values. If the quantizing intervals are unequal this cannot be done. When all quantizing intervals are the same, the term uniform quantizing is used. The term linear quantizing will be found, but this is, like military intelligence, a contradiction in terms.

The term LSB (least significant bit) will also be found in place of quantizing interval in some treatments, but this is a poor term because quantizing is not always used to create binary values and because a bit can only have two values. In studying quantizing we wish to discuss values smaller than a quantizing interval, but a fraction of an LSB is a contradiction in terms.

Whatever the exact voltage of the input signal, the quantizer will determine the quantizing interval in which it lies. In what may be considered a separate step, the quantizing interval is then allocated a code value which is typically some form of binary number. The

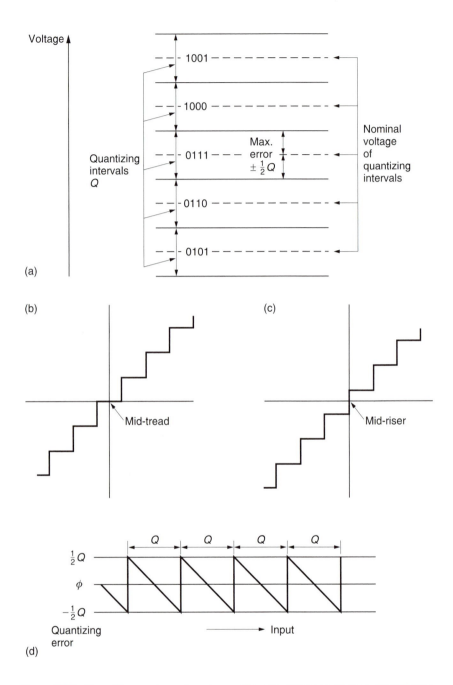

Figure 2.22 Quantizing assigns discrete numbers to variable voltages. All voltages within the same quantizing interval are assigned the same number which causes a DAC to produce the voltage at the centre of the intervals shown by the dashed lines in (a). This is the characteristic of the mid-tread quantizer shown in (b). An alternative system is the mid-riser system shown in (c). Here 0 volts analog falls between two codes and there is no code for zero. Such quantizing cannot be used prior to signal processing because the number is no longer proportional to the voltage. Quantizing error cannot exceed ± ½Q as shown in (d).

information sent is the number of the quantizing interval in which the input voltage lay. Exactly where that voltage lay within the interval is not conveyed, and this mechanism puts a limit on the accuracy of the quantizer. When the number of the quantizing interval is converted back to the analog domain, it will result in a voltage at the centre of the quantizing interval as this minimizes the magnitude of the error between input and output. The number range is limited by the wordlength of the binary numbers used. In a 16-bit system commonly used for audio, 65 536 different quantizing intervals exist, whereas video systems typically have 8-bit systems having 256 quantizing intervals.

2.14 Quantizing error

It is possible to draw a transfer function for such an ideal quantizer followed by an ideal DAC, and this is shown in Figure 2.22(b). A transfer function is simply a graph of the output with respect to the input. When the term linearity is used, this generally means the straightness of the transfer function. Linearity is a goal in audio and video, yet it will be seen that an ideal quantizer is anything but. Quantizing causes a voltage error in the sample which cannot exceed $\pm\frac{1}{2}Q$ unless the input is so large that clipping occurs.

Figure 2.22(b) shows the transfer function is somewhat like a staircase, and the voltage corresponding to audio muting or video blanking is half-way up a quantizing interval, or on the centre of a tread. This is the so-called mid-tread quantizer which is universally used in audio and video. Figure 2.22(c) shows the alternative mid-riser transfer function which causes difficulty because it does not have a code value at muting/blanking level and as a result the code value is not proportional to the signal voltage.

In studying the transfer function it is better to avoid complicating matters with the aperture effect of the DAC. For this reason it is assumed here that output samples are of negligible duration. Then impulses from the DAC can be compared with the original analog waveform and the difference will be impulses representing the quantizing error waveform. As can be seen in Figure 2.23, the quantizing error waveform can be thought of as an unwanted signal which the quantizing process adds to the perfect original. As the transfer function is non-linear, ideal quantizing can cause distortion. As a result practical digital audio devices use non-ideal quantizers to achieve linearity. The quantizing error of an ideal quantizer is a complex function, and it has been researched in great depth.[9]

As the magnitude of the quantizing error is limited, its effect can be minimized by making the signal larger. This will require more quantizing

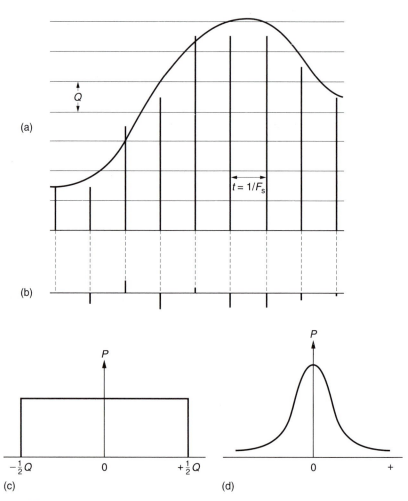

Figure 2.23 In (a) an arbitrary signal is represented to finite accuracy by PAM needles whose peaks are at the centre of the quantizing intervals. The errors caused can be thought of as an unwanted signal (b) added to the original. In (c) the amplitude of a quantizing error needle will be from $-\frac{1}{2}Q$ to $+\frac{1}{2}Q$ with equal probability. Note, however, that white noise in analog circuits generally has Gaussian amplitude distribution, shown in (d).

intervals and more bits to express them. The number of quantizing intervals multiplied by their size gives the quantizing range of the convertor. A signal outside the range will be clipped. Clearly if clipping is avoided, the larger the signal, the less will be the effect of the quantizing error.

Consider first the case where the input signal exercises the whole quantizing range and has a complex waveform. In audio this might be orchestral music; in video a bright, detailed contrasty scene. In these cases successive samples will have widely varying numerical values and the quantizing error on a given sample will be independent of that on others. In this case the size of the quantizing error will be distributed with equal probability between the limits. Figure 2.23(c) shows the resultant uniform probability density. In this case the unwanted signal added by quantizing is an additive broadband noise uncorrelated with the signal, and it is appropriate in this case to call it quantizing noise. This is not quite the same as thermal noise which has a Gaussian probability shown in Figure 2.23(d). The subjective difference is slight. Treatments which then assume that quantizing error is *always* noise give results which are at variance with reality. Such approaches only work if the probability density of the quantizing error is uniform. Unfortunately at low levels, and particularly with pure or simple waveforms, this is simply not true.

At low levels, quantizing error ceases to be random, and becomes a function of the input waveform and the quantizing structure. Once an unwanted signal becomes a deterministic function of the wanted signal, it has to be classed as a distortion rather than a noise. We predicted a distortion because of the non-linearity or staircase nature of the transfer function. With a large signal, there are so many steps involved that we must stand well back, and a staircase with many steps appears to be a slope. With a small signal there are few steps and they can no longer be ignored.

The non-linearity of the transfer function results in distortion, which produces harmonics. Unfortunately these harmonics are generated *after* the anti-aliasing filter, and so any which exceed half the sampling rate will alias. Figure 2.24 shows how this results in anharmonic distortion in audio.

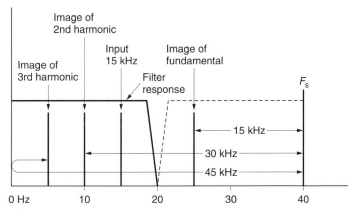

Figure 2.24 Quantizing produces distortion *after* the anti-aliasing filter: thus the distortion products will fold back to produce anharmonics in the audio hand. Here the fundamental of 15 kHz produces second and third harmonic distortion at 30 and 45 kHz. This results in aliased products at 40 − 30 = 10 kHz and 40 − 45 = (−)5 kHz.

These anharmonics result in spurious tones known as birdsinging. When the sampling rate is a multiple of the input frequency the result is harmonic distortion. Where more than one frequency is present in the input, intermodulation distortion occurs, which is known as granulation.

As the input signal is further reduced in level, it may remain within one quantizing interval. The output will be silent because the signal is now the quantizing error. In this condition, low-frequency signals such as air-conditioning rumble can shift the input in and out of a quantizing interval so that the quantizing distortion comes and goes, resulting in noise modulation.

In video, quantizing error in luminance results in visible contouring on low-key scenes or flat fields. Slowly changing brightness across the screen is replaced by areas of constant brightness separated by sudden steps. In colour difference signals, contouring results in an effect known as posterization where subtle variations in colour are removed and large areas are rendered by the same colour as if they had been painted by numbers.

2.15   Dither

At high signal level, quantizing error is effectively noise. As the level falls, the quantizing error of an ideal quantizer becomes more strongly correlated with the signal and the result is distortion. If the quantizing error can be decorrelated from the input in some way, the system can remain linear. Dither performs the job of decorrelation by making the action of the quantizer unpredictable.

The first documented use of dither was in picture coding.[10] In this system, the noise added prior to quantizing was subtracted after reconversion to analog. This is known as subtractive dither. Although subsequent subtraction has some slight advantages,[9] it suffers from practical drawbacks, since the original noise waveform must accompany the samples or must be synchronously re-created at the DAC. This is virtually impossible in a system where the signal may have been edited. Practical systems use non-subtractive dither where the dither signal is added prior to quantization and no subsequent attempt is made to remove it. The introduction of dither inevitably causes a slight reduction in the signal-to-noise ratio attainable, but this reduction is a small price to pay for the elimination of non-linearities. As linearity is an essential requirement for digital audio and video, the use of dither is equally essential.

The ideal (noiseless) quantizer of Figure 2.23 has fixed quantizing intervals and must always produce the same quantizing error from the same signal. In Figure 2.25 it can be seen that an ideal quantizer can be

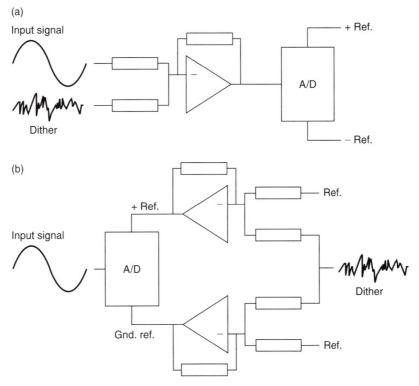

Figure 2.25 Dither can be applied to a quantizer in one of two ways. In (a) the dither is linearly added to the analog input signal, whereas in (b) it is added to the reference voltages of the quantizer.

dithered by linearly adding a controlled level of noise either to the input signal or to the reference voltage which is used to derive the quantizing intervals. There are several ways of considering how dither works, all of which are valid.

The addition of dither means that successive samples effectively find the quantizing intervals in different places on the voltage scale. The quantizing error becomes a function of the dither, rather than just a function of the input signal. The quantizing error is not eliminated, but the subjectively unacceptable distortion is converted into broadband noise which is more benign.

An alternative way of looking at dither is to consider the situation where a low-level input signal is changing slowly within a quantizing interval. Without dither, the same numerical code results, and the variations within the interval are lost. Dither has the effect of forcing the quantizer to switch between two or more states. The higher the voltage of the input signal within the interval, the more probable it becomes that the output code will take on a higher value. The lower the input voltage within the interval, the more probable it is that the output code will take the lower value. The dither has resulted in a form of duty cycle modulation, and the resolution of the system has been extended indefinitely instead of being limited by the size of the steps.

Dither can also be understood by considering the effect it has on the transfer function of the quantizer. This is normally a perfect staircase, but in the presence of dither it is smeared horizontally until with a certain minimum amplitude the average transfer function becomes straight.

The characteristics of the noise used are rather important for optimal performance, although many sub-optimal but nevertheless effective systems are in use. The main parameters of interest are the peak-to-peak amplitude, and the probability distribution of the amplitude. Triangular probability works best and this can be obtained by summing the output of two uniform probability processes.

The use of dither invalidates the conventional calculations of signal-to-noise ratio available for a given wordlength. This is of little consequence as the rule of thumb that multiplying the number of bits in the wordlength by 6 dB gives the SNR a result that will be close enough for all practical purposes.

It has only been possible to introduce the principles of conversion of audio and video signals here. For more details of the operation of convertors the reader is referred elsewhere.[1,11]

2.16   Binary codes for audio

For audio use, the prime purpose of binary numbers is to express the values of the samples which represent the original analog sound-pressure waveform. There will be a fixed number of bits in the sample, which determines the number range. In a 16-bit code there are 65 536 different numbers. Each number represents a different analog signal voltage, and care must be taken during conversion to ensure that the signal does not go outside the converter range, or it will be clipped. In Figure 2.26 it will be seen that in a simple system, the number range goes from 0000 hex, which represents the largest negative voltage, through 7FFF hex, which represents the smallest negative voltage, through 8000 hex, which represents the smallest positive voltage, to FFFF hex, which represents the largest positive voltage. Effectively, the number range of the converter has been shifted so that positive and negative voltages in a real audio signal can be expressed by binary numbers which are only positive. This approach is called offset binary, and is perfectly acceptable where the signal has been digitized only for recording or transmission from one place to another, after which it will be converted back to analog. Under these conditions it is not necessary for the quantizing steps to be uniform,

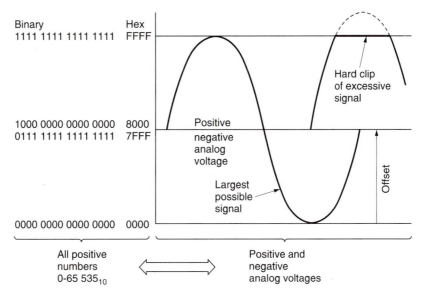

Figure 2.26 Offset binary coding is simple but causes problems in digital audio processing. It is seldom used.

provided both ADC and DAC are constructed to the same standard. In practice, it is the requirements of signal processing in the digital domain which make both non-uniform quantizing and offset binary unsuitable.

Figure 2.27 shows that an audio signal voltage is referred to midrange. The level of the signal is measured by how far the waveform deviates from

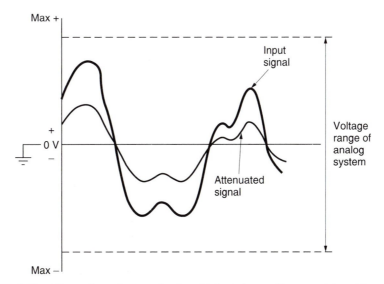

Figure 2.27 Attenuation of an audio signal takes place with respect to midrange.

midrange, and attenuation, gain and mixing all take place around midrange. It is necessary to add sample values from two or more different sources to perform the mixing function, and adding circuits assume that all bits represent the same quantizing interval so that the sum of two sample values will represent the sum of the two original analog voltages. In non-uniform quantizing this is not the case, and such signals cannot readily be processed. If two offset binary sample streams are added together in an attempt to perform digital mixing, the result will be an offset which may lead to an overflow. Similarly, if an attempt is made to attenuate by, say, 6 dB by dividing all of the sample values by two, Figure 2.28 shows that a further offset results. The problem is that offset binary is referred to one end of the range. What is needed is a numbering system which operates symmetrically about the centre of the range.

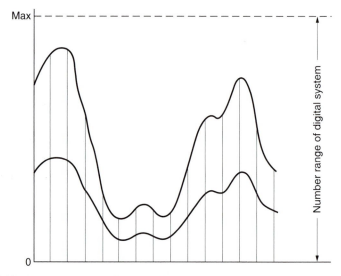

Figure 2.28 The result of an attempted attenuation in pure binary code is an offset. Pure binary cannot be used for digital audio processing.

In the two's complement system, the upper half of the pure binary number range has been redefined to represent negative quantities. If a pure binary counter is constantly incremented and allowed to overflow, it will produce all the numbers in the range permitted by the number of available bits, and these are shown for a 4-bit example drawn around the circle in Figure 2.29. As a circle has no real beginning, it is possible to consider it to start wherever it is convenient. In two's complement, the quantizing range represented by the circle of numbers does not start at zero, but starts on the diametrically opposite side of the circle. Zero is midrange, and all numbers with the MSB (most significant bit) set are considered negative. The MSB is

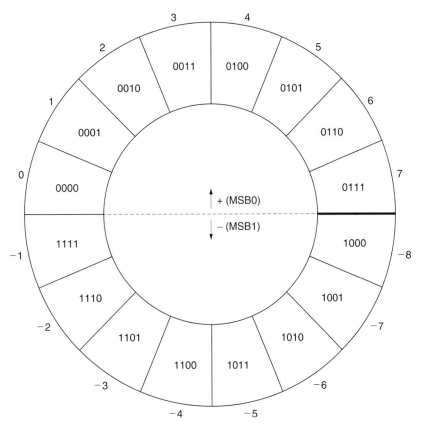

Figure 2.29 In this example of a 4-bit two's complement code, the number range is from −8 to + 7. Note that the MSB determines polarity.

thus the equivalent of a sign bit where 1 = minus. Two's complement notation differs from pure binary in that the most significant bit is inverted in order to achieve the half-circle rotation.

Figure 2.30 shows how a real ADC is configured to produce two's complement output. In (a) an analog offset voltage equal to one half the quantizing range is added to the bipolar analog signal in order to make it unipolar as in (b). The ADC produces positive only numbers in (c) which are proportional to the input voltage. The MSB is then inverted in (d) so that the all-zeros code moves to the centre of the quantizing range. The analog offset is often incorporated into the ADC as is the MSB inversion. Some convertors are designed to be used in either pure binary or two's complement mode. In this case the designer must arrange the appropriate DC conditions at the input. The MSB inversion may be selectable by an external logic level.

The two's complement system allows two sample values to be added, or mixed in audio parlance, and the result will be referred to the system

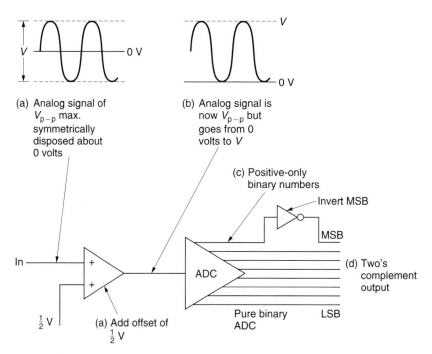

Figure 2.30 A two's complement ADC. In (a) an analog offset voltage equal to one-half the quantizing range is added to the bipolar analog signal in order to make it unipolar as in (b). The ADC produces positive-only numbers in (c), but the MSB is then inverted in (d) to give a two's complement output.

midrange; this is analogous to adding analog signals in an operational amplifier.

Figure 2.31 illustrates how adding two's complement samples simulates the audio mixing process. The waveform of input A is depicted by solid black samples, and that of B by samples with a solid outline. The result of mixing is the linear sum of the two waveforms obtained by adding pairs of sample values. The dashed lines depict the output values. Beneath each set of samples is the calculation which will be seen to give the correct result. Note that the calculations are pure binary. No special arithmetic is needed to handle two's complement numbers.

Figure 2.32 shows some audio waveforms at various levels with respect to the coding values. Where an audio waveform just fits into the quantizing range without clipping it has a level which is defined as 0 dBFs where Fs indicates *full scale*. Reducing the level by 6.02 dB makes the signal half as large and results in the second bit in the sample becoming the same as the sign bit. Reducing the level by a further 6.02 dB to −12 dBFs will make the second and third bits the same as the sign bit and so on. If a signal at −36 dBFs is input to a 16-bit system, only 10 bits will be active, the remainder will copy the sign bit. For the best

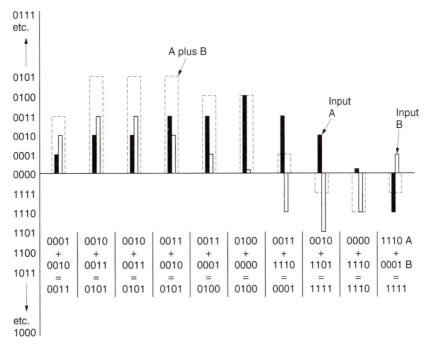

Figure 2.31 Using two's complement arithmetic, single values from two waveforms are added together with respect to midrange to give a correct mixing function.

performance, analog inputs to digital systems must have sufficient levels to exercise the whole quantizing range.

Using inversion, signal subtraction can be performed using only adding logic. The inverted input is added to perform a subtraction, just as in the analog domain. This permits a significant saving in hardware

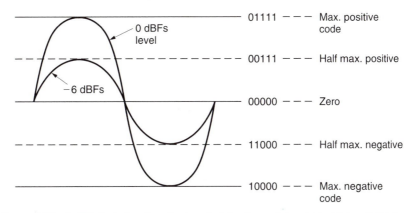

Figure 2.32 0 dBFs is defined as the level of the largest sinusoid which will fit into the quantizing range without clipping.

complexity, since only carry logic is necessary and no borrow mechanism need be supported.

In summary, two's complement notation is the most appropriate scheme for bipolar signals, and allows simple mixing in conventional binary adders. It is in virtually universal use in digital audio processing, and is accordingly adopted by all the major digital audio interfaces and recording formats. Two's complement format is also adopted for transform coding of both video and audio so that bipolar coefficients can be handled.

Two's complement numbers can have a radix point and bits below it just as pure binary numbers can. It should, however, be noted that in two's complement, if a radix point exists, numbers to the right of it are added. For example, 1100.1 is not -4.5, it is $-4 + 0.5 = -3.5$.

2.17   Binary codes for component video

In video it is important to standardize the relationship between the absolute analog voltage of the waveform and the digital code value used to express it so that all machines will interpret the numerical data in the same way. These relationships are in the voltage domain and are independent of the scanning standard used.

Figure 2.33 shows how the luminance signal fits into the quantizing range of an 8-bit system. Numbering for 10-bit systems is shown with figures for 8 bits in brackets. Black is at a level of 64_{10} (16_{10}) and peak white is at 940_{10} (235_{10}) so that there is some tolerance of imperfect analog signals. The sync pulse will clearly go outside the quantizing range, but this is of no consequence as conventional syncs are not transmitted. The visible voltage range fills the quantizing range and this gives the best possible resolution.

The colour difference signals use offset binary, where 512_{10} (128_{10}) is the equivalent of blanking voltage. The peak analog limits are reached at

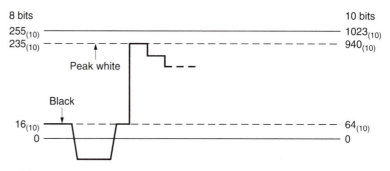

Figure 2.33   The standard luminance signal fits into 8- or 10-bit quantizing structures as shown here.

64_{10} (16_{10}) and 960_{10} (240_{10}) respectively allowing once more some latitude for maladjusted analog inputs.

Note that the code values corresponding to the eight most significant bits being all ones or all zeros (i.e. the two extreme ends of the quantizing range in 8-bit data) are not allowed to occur in the active line as they are reserved for synchronizing. Convertors must be followed by circuitry which catches these values and forces the LSB to a different value if out-of-range analog inputs are applied. The use of these reserved codes in digital interfaces precludes the use of two's complement coding, but, as was shown above, the offset is exactly half scale and so the data can be expressed in two's complement by inverting the MSB.

The peak-to-peak amplitude of Y is 880 (220) quantizing intervals, whereas for the colour difference signals it is 900 (225) intervals. There is thus a small gain difference between the signals. This will be cancelled out by the opposing gain difference at any future DAC, but must be borne in mind when digitally converting to other standards.

Although the coding standards of digital video interfaces use offset binary, for the purposes of transform coding the video must be converted to two's complement. The colour difference signals simply require the MSB to be inverted, whereas the luminance signal has an offset of half full scale subtracted first. This corresponds to 128 in 8-bit systems and 512 in 10-bit systems.

2.18   Introduction to digital processes

However complex a digital process, it can be broken down into smaller stages until finally one finds that there are really only two basic types of element in use, and these can be combined in some way and supplied with a clock to implement virtually any process. Figure 2.34 shows that the first type is a *logical* element. This produces an output which is a logical function of the input with minimal delay. The second type is a *storage* element which samples the state of the input(s) when clocked and holds or delays that state. The strength of binary logic is that the signal has only two states, and considerable noise and distortion of the binary waveform can be tolerated before the state becomes uncertain. At every logical element, the signal is compared with a threshold, and can thus can pass through any number of stages without being degraded. In addition, the use of a storage element at regular locations throughout logic circuits eliminates time variations or jitter. Figure 2.34 shows that if the inputs to a logic element change, the output will not change until the *propagation delay* of the element has elapsed. However, if the output of the logic element forms the input to a storage element, the output of that element will not change until the input is sampled *at the next clock edge*. In this way

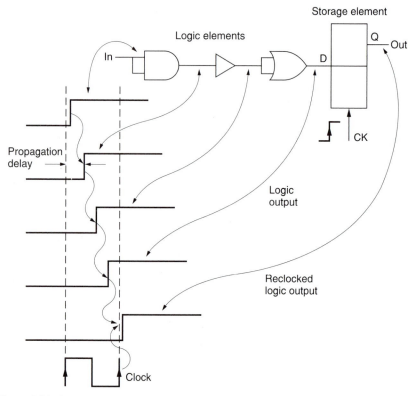

Figure 2.34 Logic elements have a finite propagation delay between input and output and cascading them delays the signal an arbitrary amount. Storage elements sample the input on a clock edge and can return a signal to near coincidence with the system clock. This is known as reclocking. Reclocking eliminates variations in propagation delay in logic elements.

the signal edge is aligned to the system clock and the propagation delay of the logic becomes irrelevant. The process is known as reclocking.

2.19   Logic elements

The two states of the signal when measured with an oscilloscope are simply two voltages, usually referred to as high and low. The actual voltage levels will depend on the type of logic family in use, and on the supply voltage used. Within logic, these levels are not of much consequence, and it is only necessary to know them when interfacing between different logic families or when driving external devices. The pure logic designer is not interested at all in these voltages, only in their meaning. Just as the electrical waveform from a microphone represents sound velocity, so the waveform in a logic circuit represents the truth of some statement. As there are only two states, there can only be *true* or *false* meanings. The true state of the signal can be assigned by the designer to either voltage state. When a high voltage represents a true logic condition and a low voltage represents a false condition, the system is known as *positive logic*, or *high true* logic. This is the usual system, but sometimes the low voltage represents the true condition and the high voltage represents the false condition. This is known as *negative logic* or *low true* logic. Provided that everyone is aware of the logic convention in use, both work equally well.

In logic systems, all logical functions, however complex, can be configured from combinations of a few fundamental logic elements or *gates*. It is not profitable to spend too much time debating which are the truly fundamental ones, since most can be made from combinations of others. Figure 2.35 shows the important simple gates and their

Positive logic name	Boolean expression	Positive logic symbol	Positive logic truth table	Plain English
Inverter or NOT gate	$Q = \bar{A}$	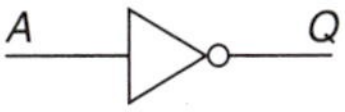	A | Q 0 | 1 1 | 0	Output is opposite of input
AND gate	$Q = A \cdot B$	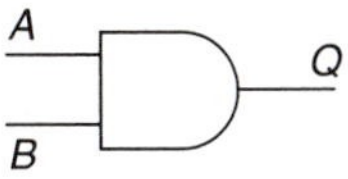	A B | Q 0 0 | 0 0 1 | 0 1 0 | 0 1 1 | 1	Output true when both inputs are true only
NAND (Not AND) gate	$Q = \overline{A \cdot B}$ $= \bar{A} + \bar{B}$	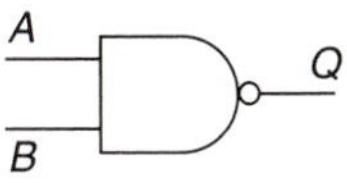	A B | Q 0 0 | 1 0 1 | 1 1 0 | 1 1 1 | 0	Output false when both inputs are true only
OR gate	$Q = A + B$	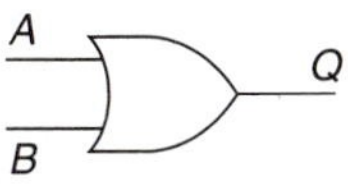	A B | Q 0 0 | 0 0 1 | 1 1 0 | 1 1 1 | 1	Output true if either or both inputs true
NOR (Not OR) gate	$Q = \overline{A + B}$ $= \bar{A} \cdot \bar{B}$	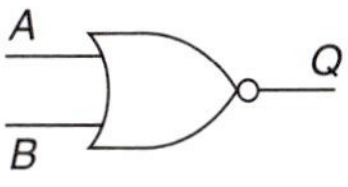	A B | Q 0 0 | 1 0 1 | 0 1 0 | 0 1 1 | 0	Output false if either or both inputs true
Exclusive OR (XOR) gate	$Q = A \oplus B$	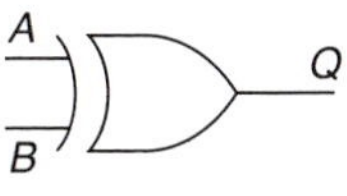	A B | Q 0 0 | 0 0 1 | 1 1 0 | 1 1 1 | 0	Output true if inputs are different

Figure 2.35 The basic logic gates compared.

derivatives, and introduces the logical expressions to describe them, which can be compared with the truth-table notation. The figure also shows the important fact that when negative logic is used, the OR gate function interchanges with that of the AND gate.

If numerical quantities need to be conveyed down the two-state signal paths described here, then the only appropriate numbering system is binary, which has only two symbols, 0 and 1. Just as positive or negative logic could be used for the truth of a logical binary signal, it can also be used for a numerical binary signal. Normally, a high voltage level will represent a binary 1 and a low voltage will represent a binary 0, described as a 'high for a one' system. Clearly a 'low for a one' system is just as feasible. Decimal numbers have several columns, each of which represents a different power of ten; in binary the column position specifies the power of two.

2.20   Storage elements

The basic memory element in logic circuits is the latch, which is constructed from two gates as shown in Figure 2.36(a), and which can be set or reset. A more useful variant is the D-type latch shown at (b) which remembers the state of the input at the time a separate clock either changes state for an edge-triggered device, or after it goes false for a level-triggered device. D-type latches are commonly available with four or eight latches to the chip. A shift register can be made from a series of latches by connecting the Q output of one latch to the D input of the next and connecting all of the clock inputs in parallel. Data are delayed by the number of stages in the register. Shift registers are also useful for converting between serial and parallel data transmissions.

Where large numbers of bits are to be stored, cross-coupled latches are less suitable because they are more complicated to fabricate inside integrated circuits than dynamic memory, and consume more current.

In large random access memories (RAMs), the data bits are stored as the presence or absence of charge in a tiny capacitor as shown in Figure 2.36(c). The capacitor is formed by a metal electrode, insulated by a layer of silicon dioxide from a semiconductor substrate, hence the term MOS (Metal Oxide Semiconductor). The charge will suffer leakage, and the value would become indeterminate after a few milliseconds. Where the delay needed is less than this, decay is of no consequence, as data will be read out before it has had a chance to decay. Where longer delays are necessary, such memories must be refreshed periodically by reading the bit value and writing it back to the same place. Most modern MOS RAM chips have suitable circuitry built in. Large RAMs store thousands of bits, and it is clearly impractical to have a connection to each one. Instead, the

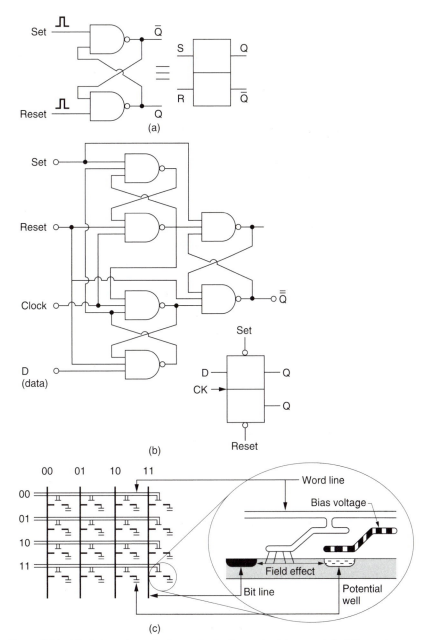

Figure 2.36 Digital semiconductor memory types. In (a), one data bit can be stored in a simple set–reset latch, which has little application because the D-type latch in (b) can store the state of the single data input when the clock occurs. These devices can be implemented with bipolar transistors of FETs, and are called static memories because they can store indefinitely. They consume a lot of power. In (c), a bit is stored as the charge in a potential well in the substrate of a chip. It is accessed by connecting the bit line with the field effect from the word line. The single well where the two lines cross can then be written or read. These devices are called dynamic RAMs because the charge decays, and they must be read and rewritten (refreshed) periodically.

desired bit has to be addressed before it can be read or written. The size of the chip package restricts the number of pins available, so that large memories use the same address pins more than once. The bits are arranged internally as rows and columns, and the row address and the column address are specified sequentially on the same pins.

2.21   Binary adding

The circuitry necessary for adding pure binary or two's complement numbers is shown in Figure 2.37. Addition in binary requires two bits to be taken at a time from the same position in each word, starting at the least significant bit. Should both be ones, the output is zero, and there is a *carry-out* generated. Such a circuit is called a half adder, shown in Figure 2.37(a) and is suitable for the least significant bit of the calculation. All higher stages will require a circuit which can accept a carry input as well as two data inputs. This is known as a full adder (Figure 2.37(b)). Multibit full adders are available in chip form, and have carry-in and carry-out terminals to allow them to be cascaded to operate on long wordlengths. Such a device is also convenient for inverting a two's complement number, in conjunction with a set of inverters. The adder chip has one set of inputs grounded, and the carry-in permanently held true, such that it adds one to the one's complement number from the inverter.

When mixing by adding sample values, care has to be taken to ensure that if the sum of the two sample values exceeds the number range the result will be clipping rather than wraparound. In two's complement, the action necessary depends on the polarities of the two signals. Clearly if one positive and one negative number are added, the result cannot exceed the number range. If two positive numbers are added, the symptom of positive overflow is that the most significant bit sets, causing an erroneous negative result, whereas a negative overflow results in the most significant bit clearing. The overflow control circuit will be designed to detect these two conditions, and override the adder output. If the MSB of both inputs is zero, the numbers are both positive, thus if the sum has the MSB set, the output is replaced with the maximum positive code (0111 ...). If the MSB of both inputs is set, the numbers are both negative, and if the sum has no MSB set, the output is replaced with the maximum negative code (1000...). These conditions can also be connected to warning indicators. Figure 2.37(c) shows this system in hardware. The resultant clipping on overload is sudden, and sometimes a PROM is included which translates values around and beyond maximum to soft-clipped values below or equal to maximum.

A storage element can be combined with an adder to obtain a number of useful functional blocks which will crop up frequently in digital equipment. Figure 2.38(a) shows that a latch is connected in a feedback

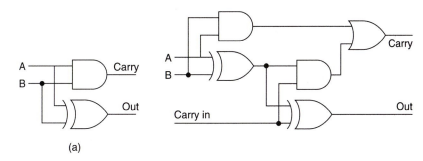

(a)

Data A	Bits B	Carry in	Out	Carry out
0	0	0	0	0
0	0	1	1	0
0	1	0	1	0
0	1	1	0	1
1	0	0	1	0
1	0	1	0	1
1	1	0	0	1
1	1	1	1	1

(b)

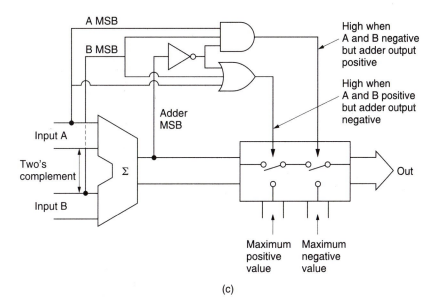

(c)

Figure 2.37 (a) Half adder; (b) full-adder circuit and truth table; (c) comparison of sign bits prevents wraparound on adder overflow by substituting clipping level.

loop around an adder. The latch contents are added to the input each time it is clocked. The configuration is known as an accumulator in computation because it adds up or accumulates values fed into it. In filtering, it is known as a discrete time integrator. If the input is held at some constant value, the output increases by that amount on each clock. The output is thus a sampled ramp.

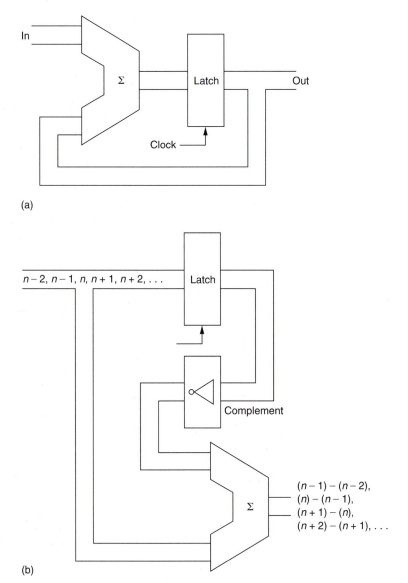

(a)

(b)

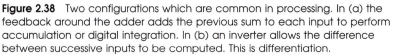

Figure 2.38 Two configurations which are common in processing. In (a) the feedback around the adder adds the previous sum to each input to perform accumulation or digital integration. In (b) an inverter allows the difference between successive inputs to be computed. This is differentiation.

Figure 2.38(b) shows that the addition of an invertor allows the difference between successive inputs to be obtained. This is digital differentiation. The output is proportional to the slope of the input.

2.22   Gain control by multiplication

Gain control is used extensively in compression systems. Digital filtering and transform calculations rely heavily on it, as do the requantizing processes which perform the actual compression. Gain is controlled in the digital domain by multiplying each sample value by a coefficient. If that coefficient is less than one, attenuation will result; if it is greater than one, amplification can be obtained.

Multiplication in binary circuits is difficult. It can be performed by repeated adding, but this is too slow to be of any use. In fast multiplication, one of the inputs will be simultaneously multiplied by one, two, four, etc., by hard-wired bit shifting. Figure 2.39 shows that the

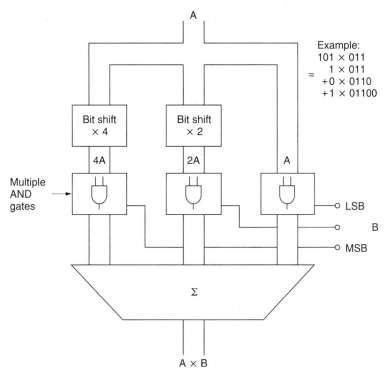

Figure 2.39 Structure of fast multiplier: the input A is multiplied by 1, 2, 4, 8, etc., by bit shifting. The digits of the B input then determine which multiples of A should be added together by enabling AND gates between the shifters and the adder. For long wordlengths, the number of gates required becomes enormous, and the device is best implemented in a chip.

other input bits will determine which of these powers will be added to produce the final sum, and which will be neglected. If multiplying by five, the process is the same as multiplying by four, multiplying by one, and adding the two products. This is achieved by adding the input to itself shifted two places. As the wordlength of such a device increases, the complexity increases exponentially, so this is a natural application for an integrated circuit.

2.23   Multiplexing principles

Multiplexing is used where several signals are to be transmitted down the same channel. The channel bit rate must be the same as or greater than the sum of the source bit rates. Figure 2.40 shows that when multiplexing is used, the data from each source has to be time compressed. This is done by buffering source data in a memory at the multiplexer. It is written into the memory in real time as it arrives, but will be read from the memory with a clock which has a much higher rate. This means that the readout occurs in a smaller timespan. If, for example, the clock frequency is raised by a factor of ten, the data for a given signal will be transmitted in a tenth of the normal time, leaving time in the multiplex for nine more such signals.

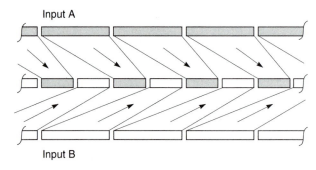

Input A

Input B

Figure 2.40 Multiplexing requires time compression on each input.

In the demultiplexer another buffer memory will be required. Only the data for the selected signal will be written into this memory at the bit rate of the multiplex. When the memory is read at the correct speed, the data will emerge with its original timebase.

In practice it is essential to have mechanisms to identify the separate signals to prevent them being mixed up and to convey the original signal clock frequency to the demultiplexer. In time-division multiplexing the timebase of the transmission is broken into equal slots, one for each signal. This makes it easy for the demultiplexer, but forces a rigid structure on all of the signals such that they must all be locked to one another and have an unchanging bit rate. Packet multiplexing overcomes these limitations.

2.24 Packets

The multiplexer must switch between different time-compressed signals to create the bitstream and this is much easier to organize if each signal is in the form of data packets of constant size. Figure 2.41 shows a packet multiplexing system.

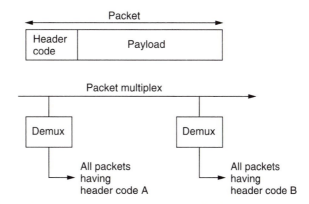

Figure 2.41 Packet multiplexing relies on headers to identify the packets.

Each packet consists of two components: the header, which identifies the packet, and the payload, which is the data to be transmitted. The header will contain at least an identification code (ID) which is unique for each signal in the multiplex. The demultiplexer checks the ID codes of all incoming packets and discards those which do not have the wanted ID.

In complex systems it is common to have a mechanism to check that packets are not lost or repeated. This is the purpose of the packet continuity count which is carried in the header. For packets carrying the same ID, the count should increase by one from one packet to the next. Upon reaching the maximum binary value, the count overflows and recommences.

2.25 Statistical multiplexing

Packet multiplexing has advantages over time-division multiplexing because it does not set the bit rate of each signal. A demultiplexer simply checks packet IDs and selects all packets with the wanted code. It will do this however frequently such packets arrive. Consequently it is practicable to have variable bit rate signals in a packet multiplex. The multiplexer has to ensure that the total bit rate does not exceed the rate of the channel, but that rate can be allocated arbitrarily between the various signals.

As a practical matter is is usually necessary to keep the bit rate of the multiplex constant. With variable rate inputs this is done by creating null packets which are generally called *stuffing* or *packing*. The headers of these packets contain an unique ID which the demultiplexer does not recognize and so these packets are discarded on arrival.

In an MPEG environment, statistical multiplexing can be extremely useful because it allows for the varying difficulty of real program material. In a multiplex of several television programs, it is unlikely that all the programs will encounter difficult material simultaneously. When one program encounters a detailed scene or frequent cuts which are hard to compress, more data rate can be allocated at the allowable expense of the remaining programs which are handling easy material.

References

1. Watkinson, J.R., *The Art of Digital Audio*, Oxford: Focal Press (1994)
2. Betts, J.A., *Signal Processing Modulation and Noise*, Chapter 6. Sevenoaks: Hodder and Stoughton (1970)
3. Meyer, J., Time correction of anti-aliasing filters used in digital audio systems. *J. Audio Eng. Soc.*, **32**, 132–137 (1984)
4. Blesser, B., Advanced A/D conversion and filtering: data conversion. In B.A. Blesser, B. Locanthi and T.G. Stockham Jr (eds), *Digital Audio*, pp. 37–53. New York: Audio Engineering Society (1983)
5. Lagadec, R., Weiss, D. and Greutmann, R., High-quality analog filters for digital audio. Presented at the 67th Audio Engineering Society Convention (New York, 1980), Preprint 1707(B-4)
6. Ishida,Y. *et al.*, A PCM digital audio processor for home use VTRs. Presented at 64th AES Convention (New York, 1979), Preprint 1528.
7. Rumsey, F.J. and Watkinson, J.R., *The Digital Interface Handbook*. Oxford: Focal Press (1995)
8. Anon., AES recommended practice for professional digital audio applications employing pulse code modulation: preferred sampling frequencies. AES5–1984 (ANSI S4.28–1984), *J. Audio Eng. Soc.*, **32**, 781–785 (1984)
9. Lipshitz, S.P. *et al.*, Quantization and dither: A theoretical survey. *J. Audio Eng. Soc.*, **40**, 355–375 (1992)
10. Roberts, L. G., Picture coding using pseudo-random noise. *IRE Trans. Inform. Theory*, **IT-8** 145–154 (1962)
11. Watkinson, J.R., *The Art of Digital Video*, Oxford: Focal Press (1994)

3

Processing for compression

Virtually all compression systems rely on some combination of the basic processes outlined in this chapter. In order to understand compression a good grasp of filtering and transforms is essential along with motion estimation for video applications. These processes only express the information in the best way for the actual compression stage, which almost exclusively begins by using requantizing to shorten the word length. In this chapter the principles of filters and transforms will be explored, along with motion estimation and requantizing. These principles will be useful background for the next two chapters.

3.1 Filters

Filtering is inseparable from digital video and audio compression. There are many parallels between analog, digital and optical filters, which this section treats as a common subject. The main difference between analog and digital filters is that in the digital domain very complex architectures can be constructed at low cost in LSI and that arithmetic calculations are not subject to component tolerance or drift. Because of the sampled nature of the signal, whatever the response at low frequencies may be, all digital channels act as low-pass filters cutting off at the Nyquist limit, or half the sampling frequency.

Filtering may modify the frequency response of a system, and/or the phase response. Every combination of frequency and phase response determines the impulse response in the time domain. Figure 3.1 shows that impulse response testing tells a great deal about a filter. In a perfect filter, all frequencies should experience the same time delay. If some groups of frequencies experience a different delay from others, there is a group-delay error. As an impulse contains an infinite spectrum, a filter

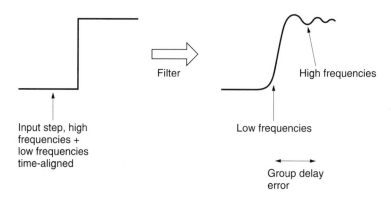

Figure 3.1   Group delay time-displaces signals as a function of frequency.

suffering from group-delay error will separate the different frequencies of an impulse along the time axis. A pure delay will cause a phase shift proportional to frequency, and a filter with this characteristic is said to be phase-linear. The impulse response of a phase-linear filter is symmetrical. If a filter suffers from group-delay error it cannot be phase-linear.

Filters can be described in two main classes, as shown in Figure 3.2, according to the nature of the impulse response. Finite-impulse response (FIR) filters are always stable and, as their name suggests, respond to an impulse once, as they have only a forward path. In the temporal domain, the time for which the filter responds to an input is finite, fixed and readily established. The same is therefore true about the distance over which a FIR filter responds in the spatial domain. FIR filters can be made perfectly phase linear if required. Most filters used for compression purposes fall into this category.

Infinite-impulse response (IIR) filters respond to an impulse indefinitely and are not necessarily stable, as they have a return path from the output to the input. For this reason they are also called recursive filters.

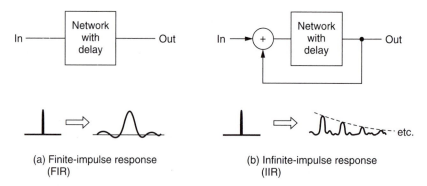

Figure 3.2   An FIR filter (a) responds only once to an input, whereas the output of an IIR filter (b) continues indefinitely rather like a decaying echo.

As the impulse response in not symmetrical, IIR filters are not phase linear. Compression systems are not intended to change the phase characteristics of the signals they handle, and so linear phase filtering is essential and only FIR filters will be considered in detail here.

A FIR filter works by graphically constructing the impulse response for every input sample. It is first necessary to establish the correct impulse response. Figure 3.3(a) shows an example of a low-pass filter which cuts off at $\frac{1}{4}$ of the sampling rate. The impulse response of a perfect low-pass filter is a $\sin x/x$ curve, where the time between the two central zero crossings is the reciprocal of the cut-off frequency. According to the mathematics, the waveform has always existed, and carries on for ever. The peak value of the output coincides with the input impulse. This means that the filter is not causal, because the output has changed before the input is known. Thus in all practical applications it is necessary to truncate the extreme ends of the impulse response, which causes an aperture effect, and to introduce a time delay in the filter equal to half the duration of the truncated impulse in order to make the filter causal. As an input impulse is shifted through the series of registers in Figure 3.3(b), the impulse response is created, because at each point it is multiplied by a coefficient as in Figure 3.3(c). These coefficients are simply the result of sampling and quantizing the desired impulse response. Clearly the sampling rate used to sample the impulse must be the same as the sampling rate for which the filter is being designed. In practice the coefficients are calculated, rather than attempting to sample an actual impulse response. The coefficient wordlength will be a compromise between cost and performance. An exception to this is in some video filters where the wordlength is short enough to allow the multipliers to be implemented as look-up tables.

Because the input sample shifts across the system registers to create the shape of the impulse response, the configuration is also known as a transversal filter. In operation with real sample streams, there will be several consecutive sample values in the filter registers at any time in order to convolve the input with the impulse response.

Simply truncating the impulse response causes an abrupt transition from input samples which matter and those which do not. Truncating the filter superimposes a rectangular shape on the time-domain impulse response. In the frequency domain the rectangular shape transforms to a $\sin x/x$ characteristic which is superimposed on the desired frequency response as a ripple. One consequence of this is known as Gibb's phenomenon; a tendency for the response to peak just before the cut-off frequency.[1,2] The shorter the window, the shallower will be the transition band of the filter. As a result, the length of the window which must be considered will depend not only on the frequency response, but also on the amount of ripple which can be tolerated.

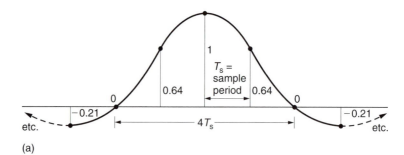

(a)

Figure 3.3(a) The impulse response of an LPF is a sin *x/x* curve which stretches from −∞ to +∞ in time. The ends of the response must be neglected, and a delay introduced to make the filter causal.

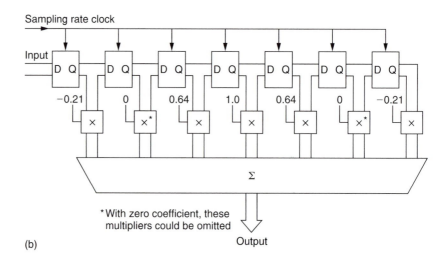

(b)

Figure 3.3(b) The structure of an FIR LPF. Input samples shift across the register and at each point are multiplied by different coefficients.

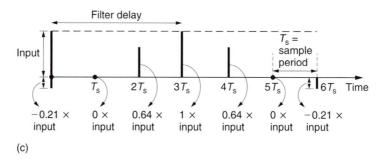

(c)

Figure 3.3(c) When a single unit sample shifts across the circuit of Figure 3.3(b), the impulse response is created at the output as the impulse is multiplied by each coefficient in turn.

If the relevant period of the impulse is measured in sample periods, the result will be the number of points or multiplications needed in the filter. Figure 3.4 compares the performance of filters with different numbers of points. General-purpose video filters such as those used in DVEs (digital video effects units) may use as few as eight points whereas a high-quality digital audio FIR filter may need as many as 96 points.

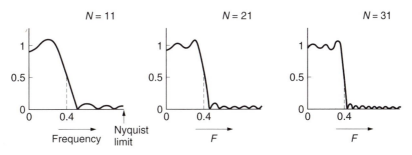

Figure 3.4 The truncation of the impulse in an FIR filter caused by the use of a finite number of points (*N*) results in ripple in the response. Shown here are three different numbers of points for the same impulse response. The filter is an LPF which rolls off at 0.4 of the fundamental interval. (Courtesy *Philips Technical Review*)

Rather than simply truncate the impulse response in time, it is better to make a smooth transition from samples which do not count to those that do. This can be done by multiplying the coefficients in the filter by a window function which peaks in the centre of the impulse. Figure 3.5 shows some different window functions and their responses. The rectangular window is the case of truncation, and the response is shown at I. A linear reduction in weight from the centre of the window to the edges characterizes the Bartlett window II. At III is shown the Hann window, which is essentially a raised cosine shape. Not shown is the similar Hamming window, which offers a slightly different trade-off between ripple and the width of the main lobe. The Blackman window introduces an extra cosine term into the Hamming window at half the period of the main cosine period, reducing Gibb's phenomenon and ripple level, but increasing the width of the transition region. The Kaiser window is a family of windows based on the Bessel function, allowing various trade-offs between ripple ratio and main lobe width. Two of these are shown in IV and V.

Filter coefficients can be optimized by computer simulation. One of the best-known techniques used is the Remez exchange algorithm, which converges on the optimum coefficients after a number of iterations.

If the coefficients are not quantized finely enough, it will be as if they had been calculated inaccurately, and the performance of the filter will

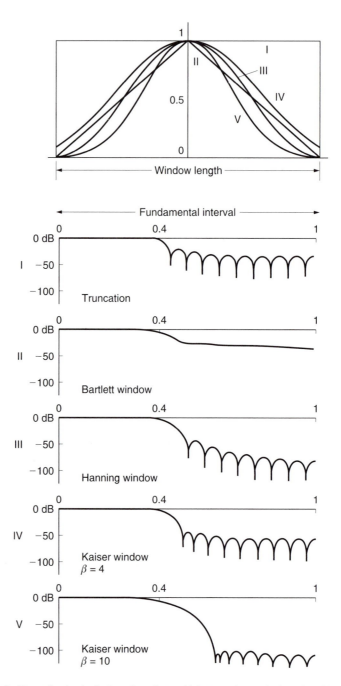

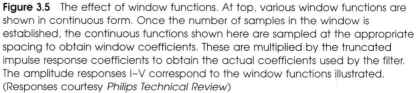

Figure 3.5 The effect of window functions. At top, various window functions are shown in continuous form. Once the number of samples in the window is established, the continuous functions shown here are sampled at the appropriate spacing to obtain window coefficients. These are multiplied by the truncated impulse response coefficients to obtain the actual coefficients used by the filter. The amplitude responses I–V correspond to the window functions illustrated. (Responses courtesy *Philips Technical Review*)

be less than expected. Figure 3.6 shows an example of quantizing coefficients. Conversely, raising the wordlength of the coefficients increases cost.

The FIR structure is inherently phase linear because it is easy to make the impulse response absolutely symmetrical. Because of this inherent phase-linearity, a FIR filter can be designed for a specific impulse response, and the frequency response will follow.

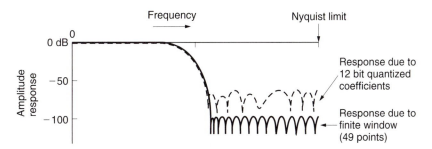

Figure 3.6 Frequency response of a 49 point transversal filter with infinite precision (solid line) shows ripple due to finite window size. Quantizing coefficients to 12 bits reduces attenuation in the stopband. (Responses courtesy *Philips Technical Review*)

The frequency response of the filter can be changed at will by changing the coefficients. A programmable filter only requires a series of PROMs to supply the coefficients; the address supplied to the PROMs will select the response. The frequency response of a digital filter will also change if the clock rate is changed, so it is often less ambiguous to specify a frequency of interest in a digital filter in terms of a fraction of the fundamental interval rather than in absolute terms. The configuration shown in Figure 3.3 serves to illustrate the principle. The units used on the diagrams are sample periods and the response is proportional to these periods or spacings, and so it is not necessary to use actual figures.

Where the impulse response is symmetrical, it is often possible to reduce the number of multiplications, because the same product can be used twice, at equal distances before and after the centre of the window. This is known as folding the filter. A folded filter is shown in Figure 3.7.

3.2   Downsampling filters

In compression pre-processing, it is often necessary to downsample the input picture. This may be needed to obtain the desired picture size, e.g. 360 pixels across, from a standard definition picture which is 720 pixels

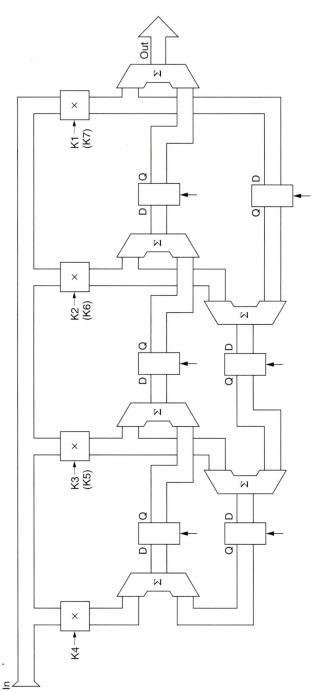

Figure 3.7 A seven-point folded filter for a symmetrical impulse reponse. In this case K1 and K7 will be identical, and the input sample can be multipled once, and the product fed into the output shift system in two different places. The centre coefficient K4 appears once. In an even-numbered filter the centre coefficient would also be used twice.

across. 4:2:2 inputs will need vertical filtering to produce 4:2:0 which most of the MPEG-2 profiles use.

Figure 3.8 shows that when a filter is used for downsampling, the stopband rejection is important because poor performance here results in aliasing. In an MPEG environment aliasing is particularly undesirable on an input signal as after the DCT stage aliasing will result in spurious coefficients which will require a higher bit rate to convey. If this bit rate is not available the picture quality will be impaired. Consequently the performance criteria for MPEG downsampling filters is more stringent than for general use and filters will generally require more points than in other applications.

In the example of Figure 3.9, a low-pass filter is shown which is intended to allow downsampling by a factor of two. The key feature is that the stopband must have begun before one half of the output sampling rate. This is most readily achieved using a Hamming window

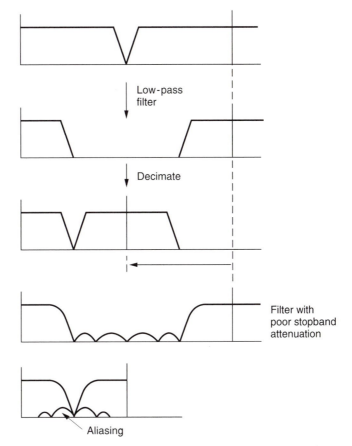

Figure 3.8 Downsampling filters for MPEG prefiltering must have good stopband rejection to avoid aliasing when the sampling rate is reduced. Stopband performance is critical for a pre-processing filter.

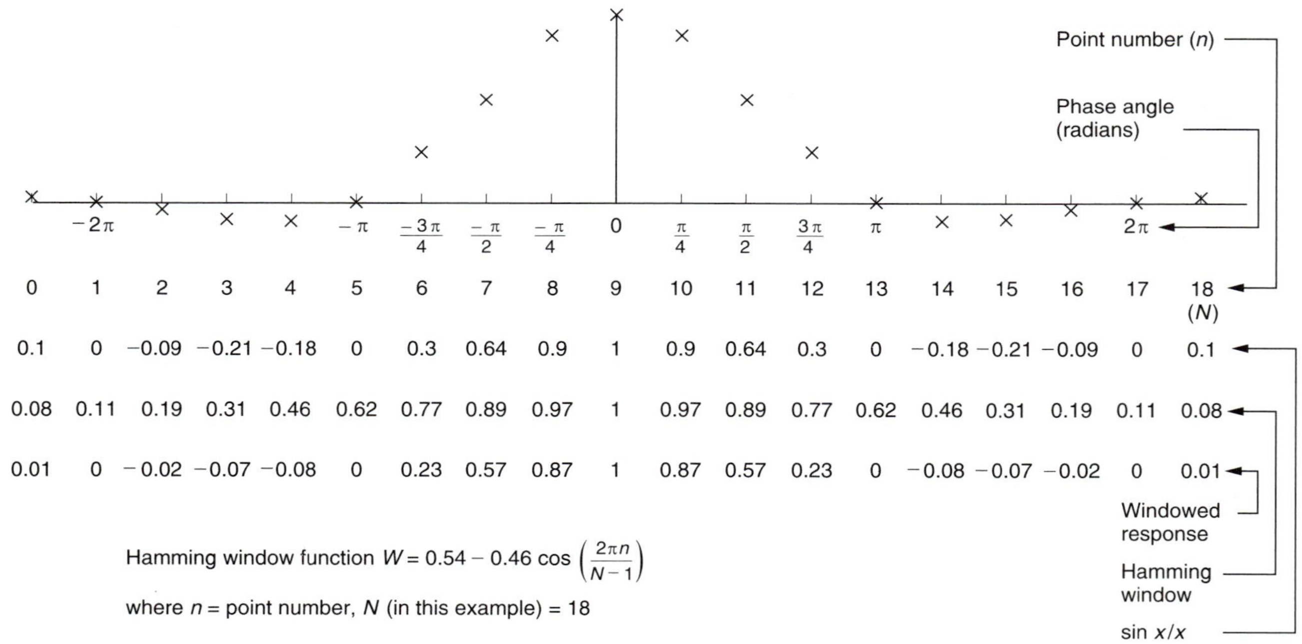

Point number (n)																		
0	1	2	3	4	5	6	7	8	9	10	11	12	13	14	15	16	17	18 (N)
Phase angle (radians)																		
	-2π				$-\pi$	$-\frac{3\pi}{4}$	$-\frac{\pi}{2}$	$-\frac{\pi}{4}$	0	$\frac{\pi}{4}$	$\frac{\pi}{2}$	$\frac{3\pi}{4}$	π				2π	
0.1	0	−0.09	−0.21	−0.18	0	0.3	0.64	0.9	1	0.9	0.64	0.3	0	−0.18	−0.21	−0.09	0	0.1
0.08	0.11	0.19	0.31	0.46	0.62	0.77	0.89	0.97	1	0.97	0.89	0.77	0.62	0.46	0.31	0.19	0.11	0.08
0.01	0	−0.02	−0.07	−0.08	0	0.23	0.57	0.87	1	0.87	0.57	0.23	0	−0.08	−0.07	−0.02	0	0.01

Windowed response
Hamming window
sin x/x

Hamming window function $W = 0.54 - 0.46 \cos\left(\frac{2\pi n}{N-1}\right)$

where n = point number, N (in this example) = 18

Figure 3.9 A downsampling filter using the Hamming window.

because it was designed empirically to have a flat stopband so that good aliasing attenuation is possible. The width of the transition band determines the number of significant sample periods embraced by the impulse. The Hamming window doubles the width of the transition band. This determines in turn both the number of points in the filter, and the filter delay. As the impulse is symmetrical, the delay will be half the impulse period. The impulse response is a $\sin x/x$ function, and this has been calculated in the figure. The equation for the Hamming window function is shown with the window values which result. The $\sin x/x$ response is next multiplied by the window function to give the windowed impulse response shown.

3.3 The quadrature mirror filter

Audio compression often uses a process known as band splitting which splits up the audio spectrum into a series of frequency ranges. Band splitting is complex and requires a lot of computation. One bandsplitting method which is useful is quadrature mirror filtering.[3] The QMF is is a kind of twin FIR filter which converts a PCM sample stream into two sample streams of half the input sampling rate, so that the output data rate equals the input data rate. The frequencies in the lower half of the audio spectrum are carried in one sample stream, and the frequencies in the upper half of the spectrum are carried in the other. While the lower frequency output is a PCM band-limited representation of the input waveform, the upper frequency output isn't.

A moment's thought will reveal that it could not be because the sampling rate is not high enough. In fact the upper half of the input spectrum has been heterodyned down to the same frequency band as the lower half by the clever use of aliasing. The waveform is unrecognizable, but when heterodyned back to its correct place in the spectrum in an inverse step, the correct waveform will result once more. Figure 3.10 shows how the idea works.

Sampling theory states that the sampling rate needed must be at least twice the bandwidth in the signal to be sampled. If the signal is band limited, the sampling rate need only be more than twice the signal *bandwidth* not the signal *frequency*. Downsampled signals of this kind can be reconstructed by a reconstruction or *synthesis* filter having a bandpass response rather than a low-pass response. As only signals within the passband can be output, it is clear from Figure 3.10 that the waveform which will result is the original as the intermediate aliased waveform lies outside the passband.

Figure 3.11 shows the operation of a simple QMF. In (a) the input spectrum of the PCM audio is shown, having an audio baseband

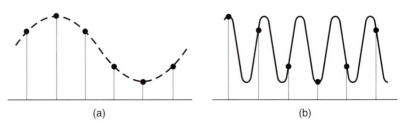

Figure 3.10 The sample stream shown would ordinarily represent the waveform shown in (a), but if it is known that the original signal could exist only between two frequencies then the waveform in (b) must be the correct one. A suitable bandpass reconstruction filter, or synthesis filter, will produce the waveform in (b).

extending up to half the sampling rate and the usual lower sideband extending down from there up to the sampling frequency. The input is passed through a FIR low-pass filter which cuts off at one quarter of the sampling rate to give the spectrum shown in (b). The input also passes in parallel through a second FIR filter which is physically identical, but the coefficients are different. The impulse response of the FIR LPF is multiplied by a cosinusoidal waveform which amplitude modulates it. The resultant impulse gives the filter a frequency response shown in (c). This is a mirror image of the LPF response. If certain criteria are met, the overall frequency response of the two filters is flat. The spectra of both (b) and (c) show that both are oversampled by a factor of 2 because they are half empty. As a result both can be decimated by a factor of two, which is the equivalent of dropping every other sample. In the case of the lower half of the spectrum, nothing remarkable happens. In the case of the upper half of the spectrum, it has been resampled at half the original frequency as shown in (d). The result is that the upper half of the audio spectrum aliases or heterodynes to the lower half.

An inverse QMF will recombine the bands into the original broadband signal. It is a feature of a QMF/inverse QMF pair that any energy near the band edge which appears in both bands due to inadequate selectivity in the filtering reappears at the correct frequency in the inverse filtering process provided that there is uniform quantizing in all the sub-bands. In practical coders, this criterion is not met, but any residual artifacts are sufficiently small to be masked.

The audio band can be split into as many bands as required by cascading QMFs in a tree. However, each stage can only divide the input spectrum in half. In some coders certain sub-bands will have passed through one splitting stage more than others and will have half their bandwidth.[4] A delay is required in the wider sub-band data for time alignment.

A simple quadrature mirror is computationally intensive because sample values are calculated which are later decimated or discarded, and

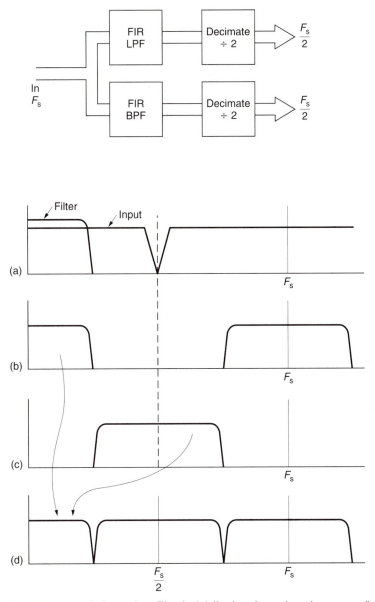

Figure 3.11 The quadrature mirror filter. In (a) the input spectrum has an audio baseband extending up to half the sampling rate. The input is passed through an FIR low-pass filter which cuts off at one-quarter of the sampling rate to give the spectrum shown in (b). The input also passes in parallel through a second FIR filter whose impulse response has been multiplied by a cosinusoidal waveform in order to amplitude-modulate it. The resultant impulse gives the filter a mirror image frequency response shown in (c). The spectra of both (b) and (c) show that both are oversampled by a factor of two because they are half empty. As a result both can be decimated by a factor of two, resulting in (d) in two identical Nyquist-sampled frequency bands of half the original width.

an alternative is to use polyphase pseudo-QMF filters[5] or wave filters[6] in which the filtering and decimation process is combined. Only wanted sample values are computed. In a polyphase filter a set of samples is shifted into position in the transversal register and then these are multiplied by different sets of coefficients and accumulated in each of several phases to give the value of a number of different samples between input samples. In a polyphase QMF, the same approach is used. Figure 3.12 shows an example of a 32-band polyphase QMF having a 512-sample window. With 32 sub-bands, each band will be decimated to $\frac{1}{32}$ of the input sampling rate. Thus only one sample in 32 will be retained after the combined filter/ decimate operation. The polyphase QMF only computes the value of the

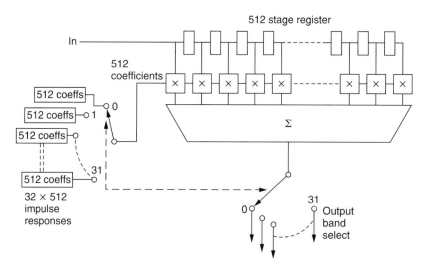

Figure 3.12 In polyphase QMF the same input samples are subject to computation using coefficient sets in many different time-multiplexed phases. The decimation is combined with the filtering so only wanted values are computed.

sample which is to be retained in each sub-band. The filter works in 32 different phases with the same samples in the transversal register. In the first phase, the coefficients will describe the impulse response of a low-pass filter, the so-called prototype filter, and the result of 512 multiplications will be accumulated to give a single sample in the first band. In the second phase the coefficients will be obtained by multiplying the impulse response of the prototype filter by a cosinusoid at the centre frequency of the second band. Once more 512 multiply accumulates will be required to obtain a single sample in the second band. This is repeated for each of the 32 bands, and in each case a different centre frequency is obtained by multiplying the prototype impulse by a different modulating frequency. Following 32 such computations, 32 output samples, one in each band, will have been computed. The transversal register then shifts 32 samples and the process repeats.

The principle of the polyphase QMF is not so different from the techniques used to compute a frequency transform and effectively blurs the distinction between sub-band coding and transform coding.

3.4 Filtering for video noise reduction

The basic principle of all video noise reducers is that there is a certain amount of correlation between the video content of successive frames, whereas there is no correlation between the noise content.

A basic recursive device is shown in Figure 3.13. There is a frame store which acts as a delay, and the output of the delay can be fed back to the input through an attenuator, which in the digital domain will be a multiplyer. In the case of a still picture, successive frames will be identical, and the recursion will be large. This means that the output video will actually be the average of many frames. If there is movement of the image, it will be necessary to reduce the amount of recursion to prevent the generation of trails or smears. Probably the most famous examples of recursion smear are the television pictures sent back of astronauts walking on the moon. The received pictures were very noisy and needed a lot of averaging to make them viewable. This was fine until the astronaut moved. The technology of the day did not permit motion sensing.

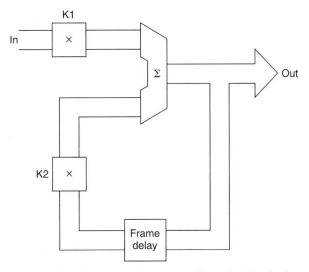

Figure 3.13 A basic recursive device feeds back the output to the input via a frame store which acts as a delay. The characteristics of the device are controlled totally by the values of the two coefficients K1 and K2 which control the multipliers.

The noise reduction increases with the number of frames over which the noise is integrated, but image motion prevents simple combining of frames. If motion estimation is available, the image of a moving object in a particular frame can be integrated from the images in several frames which have been superimposed on the same part of the screen by displacements derived from the motion measurement. The result is that greater reduction of noise becomes possible.[7]

In a median filter, sample values adjacent to the one under examination are considered. These may be in the same place in previous or subsequent images, or nearby in the same image. A median filter computes the distribution of values on all its input points. If the value of the centre point lies centrally within the distribution then it is considered to be valid and is passed to the output without change. In this case the median filter has no effect whatsoever. However, if the value of the centre point is at the edge of the distribution it is considered to be in error due to impulsive noise and a different input point, nearest the mean, is selected as the output pixel. Effectively a pixel value from nearby is used to conceal the error. The median filter is very effective against dropouts, film dirt and bit errors.

3.5   Transforms

At its simplest a transform is a process which takes information in one domain and expresses it in another. Audio and video signals are in the time domain: their voltages (or sample values) change as a function of time. Such signals are often transformed to the frequency domain for the purposes of compression. While frequency in audio has traditionally meant temporal frequency measured in Hertz, frequency in optics can also be spatial and measured in lines per millimetre (mm^{-1}) or in television in cycles per picture height (cph) or width (cpw). In the frequency domain the signal has been converted to a spectrum; a table of the energy at different temporal or spatial frequencies. If the input is repeating, the spectrum will also be sampled; i.e. it will exist at discrete frequencies. In real program material, all frequencies are seldom present together and so those which are absent need not be transmitted and a coding gain is obtained. On reception an inverse transform or synthesis process converts back from the frequency domain to the time or space domains. To take a simple analogy, a frequency transform of piano music effectively works out what frequencies are present as a function of time; the transform works out which notes were played and so the information is in a similar form to that contained in the original sheet music.

A common way of entering the frequency domain from the time or spatial domains is the Fourier transform or its equivalent in sampled systems, the discrete Fourier transform (DFT). Fourier analysis holds that any periodic waveform can be reproduced by adding together an arbitrary number of harmonically related sinusoids of various amplitudes and phases. Figure 3.14 shows how a square wave can be built up of harmonics. The spectrum can be drawn by plotting the amplitude of the harmonics against frequency. It will be seen that this gives a spectrum which is a decaying wave. It passes through zero at all even multiples of the fundamental. The shape of the spectrum is a $\sin x/x$ curve. If a square wave has a $\sin x/x$ spectrum, it follows that a filter with a rectangular impulse response will have a $\sin x/x$ frequency response.

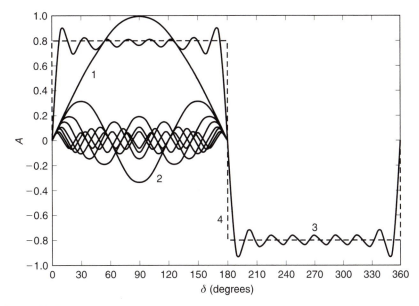

Figure 3.14 Fourier analysis of a square wave into fundamental and harmonics. A. amplitude; δ, phase of fundamental wave in degrees; 1, first harmonic (fundamental); 2, odd harmonics 3–15; 3, sum of harmonics 1–15; 4, ideal square wave.

It will be recalled that an ideal low-pass filter has a rectangular spectrum, and this has a $\sin x/x$ impulse response. These characteristics are known as a transform pair. In transform pairs, if one domain has one shape of the pair, the other domain will have the other shape. Thus a square wave has a $\sin x/x$ spectrum and a $\sin x/x$ impulse has a square spectrum. Figure 3.15 shows a number of transform pairs. Note the pulse pair. A time domain pulse of infinitely short duration has a flat spectrum. Thus a flat waveform, i.e. a constant voltage, has only zero Hz in its spectrum. Interestingly the transform of a Gaussian response in still Gaussian.

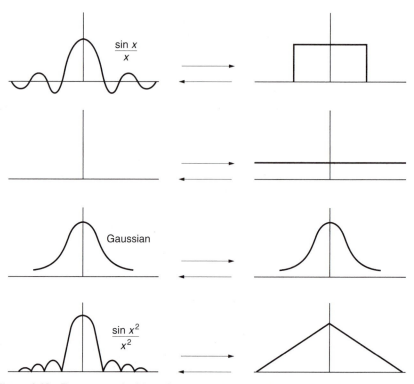

Figure 3.15 The concept of transform pairs illustrates the duality of the frequency (including spatial frequency) and time domains.

The Fourier transform specifies the amplitude and phase of the frequency components just once and such sine waves are endless. As a result the Fourier transform is only valid for periodic waveforms; i.e. those which repeat endlessly. Real program material is not like that and so it is necessary to break up the continuous time domain using windows. Figure 3.16(a) shows how a block of time is cut from the continuous input. By wrapping it into a ring it can be made to appear like a continuous periodic waveform for which a single transform, known as the short-time Fourier transform (STFT) can be computed. Note that in the Fourier transform of a periodic waveform the frequency bands have constant width. The inverse transform produces just such an endless waveform, but a window is taken from it and used as part of the output waveform. Rectangular windows are used in video compression, but are not generally adequate for audio because the discontinuities at the boundaries are audible. This can be overcome by shaping and overlapping the windows so that a cross-fade occurs at the boundaries between them as in Figure 3.16(b).

As has been mentioned, the theory of transforms assumes endless periodic waveforms. If an infinite length of waveform is available,

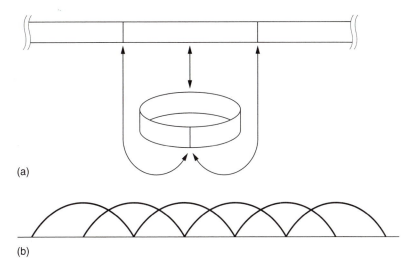

(a)

(b)

Figure 3.16 In (a) a continuous audio signal can be cut into blocks which are wrapped to make them appear periodic for the purposes of the Fourier transform. A better approach is to use overlapping windows to avoid discontinuities as in (b).

spectral analysis can be performed to infinite resolution but as the size of the window reduces, so too does the resolution of the frequency analysis. Intuitively it is clear that discrimination between two adjacent frequencies is easier if more cycles of both are available. In sampled systems, reducing the window size reduces the number of samples and so must reduce the number of discrete frequencies in the transform. Thus for good frequency resolution the window should be as large as possible. However, with large windows the time between updates of the spectrum is longer and so it is harder to locate events on the time axis. Figure 3.17(a) shows the effect of two window sizes in a conventional STFT and illustrates the principle of *uncertainty* also known as the Heisenberg inequality.

According to the uncertainty theory one can trade off time resolution against frequency resolution. In most program material, the time resolution required falls with frequency whereas the time (or spatial) resolution required rises with frequency. Compression systems using transforms sometimes split the signal into a number of frequency bands in which different window sizes are available as in Figure 3.17(b). Some have variable-length windows which are selected according to the program material. Stationary material such as steady tones are transformed with long windows whereas transients are transformed with short windows.

The recently developed wavelet transform is one in which the window length is inversely proportional to the frequency. This automatically gives

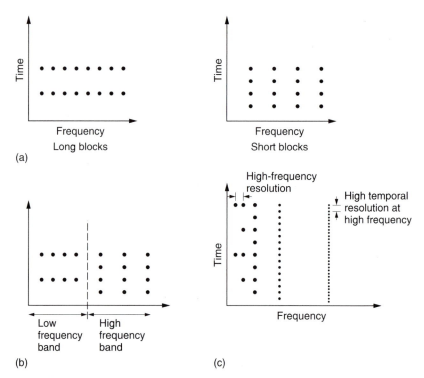

Figure 3.17   (a) In transforms greater certainty in the time domain leads to less certainty in the frequency domain and vice versa. Some transform coders split the spectrum as in (b) and use different window lengths in the two bands. In the recently developed wavelet transform the window length is inversely proportional to the frequency, giving the advantageous time/frequency characteristic shown in (c).

the advantageous time/frequency resolution characteristic shown in Figure 3.17(c).

Although compression uses transforms, the transform itself does not result in any data reduction, as there are usually as many coefficients as input samples. Paradoxically the transform increases the amount of data because the coefficient multiplications result in wordlength extension. Thus it is incorrect to refer to transform compression; instead the term transform-*based* compression should be used.

3.6   The Fourier transform

Figure 3.14 showed that if the amplitude and phase of each frequency component is known, linearly adding the resultant components in an inverse transform results in the original waveform. In digital systems the waveform is expressed as a number of discrete samples. As a result the Fourier transform analyses the signal into an equal number of discrete frequencies. This is known as a discrete Fourier transform or DFT in which the number of frequency coefficients is equal to the number of input samples. The fast Fourier transform is no more than an efficient way of computing the DFT.[8] As was seen in the previous section, practical systems must use windowing to create short-term transforms.

It will be evident from Figure 3.14 that the knowledge of the phase of the frequency component is vital, as changing the phase of any component will seriously alter the reconstructed waveform. Thus the DFT must accurately analyse the phase of the signal components.

There are a number of ways of expressing phase. Figure 3.18 shows a point which is rotating about a fixed axis at constant speed. Looked at

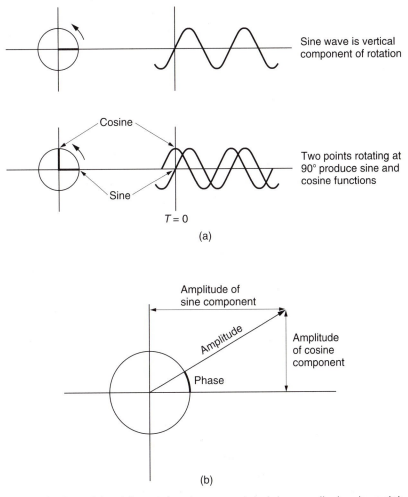

Figure 3.18 The origin of sine and cosine waves is to take a particular viewpoint of a rotation. Any phase can be synthesized by adding proportions of sine and cosine waves.

from the side, the point oscillates up and down at constant frequency. The waveform of that motion is a sine wave, and that is what we would see if the rotating point were to translate along its axis while we continued to look from the side.

One way of defining the phase of a waveform is to specify the angle through which the point has rotated at time zero ($T = 0$). If a second point is made to revolve at 90° to the first, it would produce a cosine wave when translated. It is possible to produce a waveform having arbitrary phase by adding together the sine and cosine wave in various proportions and polarities. For example, adding the sine and cosine waves in equal proportion results in a waveform lagging the sine wave by 45°.

Figure 3.18 shows that the proportions necessary are respectively the sine and the cosine of the phase angle. Thus the two methods of describing phase can be readily interchanged.

The discrete Fourier transform spectrum analyses a string of samples by searching separately for each discrete target frequency. It does this by multiplying the input waveform by a sine wave, known as the basis function, having the target frequency and adding up or integrating the products. Figure 3.19(a) shows that multiplying by basis functions gives a non-zero integral when the input frequency is the same, whereas Figure 3.19(b) shows that with a different input frequency (in fact all other different frequencies) the integral is zero showing that no component of the target frequency exists. Thus from a real waveform containing many frequencies all frequencies except the target frequency are excluded. The magnitude of the integral is proportional to the amplitude of the target component.

Figure 3.19(c) shows that the target frequency will not be detected if it is phase shifted 90° as the product of quadrature waveforms is always zero. Thus the discrete Fourier transform must make a further search for the target frequency using a cosine basis function. It follows from the arguments above that the relative proportions of the sine and cosine integrals reveal the phase of the input component. Thus each discrete frequency in the spectrum must be the result of a pair of quadrature searches.

Searching for one frequency at a time as above will result in a DFT, but only after considerable computation. However, a lot of the calculations are repeated many times over in different searches. The fast Fourier transform gives the same result with less computation by logically gathering together all the places where the same calculation is needed and making the calculation once.

The amount of computation can be reduced by performing the sine and cosine component searches together. Another saving is obtained by noting that every 180° the sine and cosine have the same magnitude but

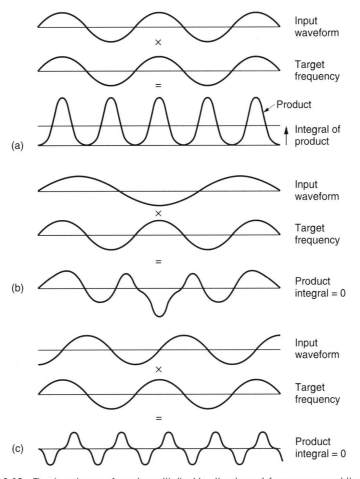

Figure 3.19 The input waveform is multiplied by the target frequency and the result is averaged or integrated. In (a) the target frequency is present and a large integral results. With another input frequency the integral is zero as in (b). The correct frequency will also result in a zero integral shown in (c) if it is at 90° to the phase of the search frequency. This is overcome by making two searches in quadrature.

are simply inverted in sign. Instead of performing four multiplications on two samples 180° apart and adding the pairs of products it is more economical to subtract the sample values and multiply twice, once by a sine value and once by a cosine value.

The first coefficient is the arithmetic mean which is the sum of all of the sample values in the block divided by the number of samples. Figure 3.20 shows how the search for the lowest frequency in a block is performed. Pairs of samples are subtracted as shown, and each difference is then multiplied by the sine and the cosine of the search

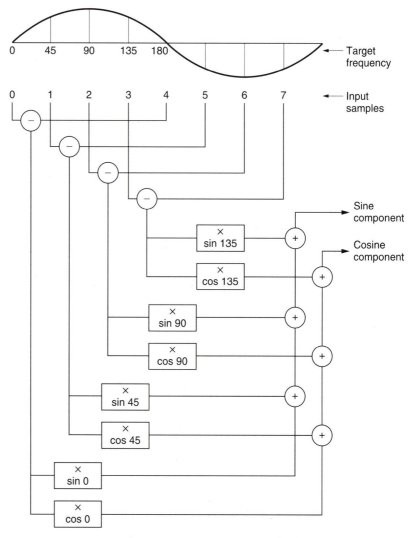

Figure 3.20 An example of a filtering search. Pairs of samples are subtracted and multiplied by sampled sine and cosine waves. The products are added to give the sine and cosine components of the search frequency.

frequency. The process shifts one sample period, and a new sample pair are subtracted and multiplied by new sine and cosine factors. This is repeated until all the sample pairs have been multiplied. The sine and cosine products are then added to give the value of the sine and cosine coefficients respectively.

It is possible to combine the calculation of the DC component which requires the sum of samples and the calculation of the fundamental which requires sample differences by combining stages shown in Figure

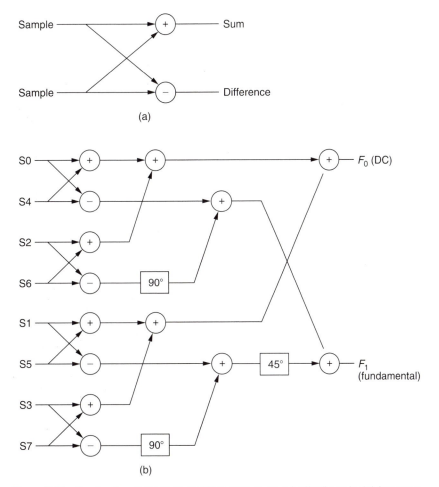

Figure 3.21 The basic element of an FFT is known as a butterfly as in (a) because of the shape of the signal paths in a sum and difference system. The use of butterflies to compute the first two coefficients is shown in (b). An actual example is given in (c) which should be compared with the result of (d) with a quadrature input. In (e) the butterflies for the first two coefficients form the basis of the computation of the third coefficient.

3.21(a) which take a pair of samples and add and subtract them. Such a stage is called a butterfly because of the shape of the schematic. Figure 3.21(b) shows how the first two components are calculated. The phase rotation boxes attribute the input to the sine or cosine component outputs according to the phase angle. As shown, the box labelled $90°$ attributes nothing to the sine output, but unity gain to the cosine output. The $45°$ box attributes the input equally to both components.

Figure 3.21(c) shows a numerical example. If a sinewave input is considered where $0°$ coincides with the first sample, this will produce

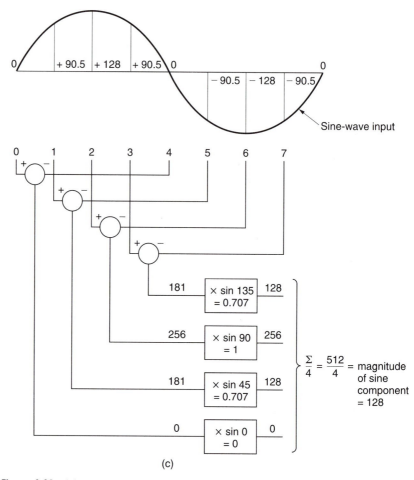

(c)

Figure 3.21 (c)

a zero sine coefficient and non-zero cosine coefficient. Figure 3.21(d) shows the same input waveform shifted by 90°. Note how the coefficients change over.

Figure 3.21(e) shows how the next frequency coefficient is computed. Note that exactly the same first-stage butterfly outputs are used, reducing the computation needed.

A similar process may be followed to obtain the sine and cosine coefficients of the remaining frequencies. The full FFT diagram for eight samples is shown in Figure 3.22(a). The spectrum this calculates is shown in (b). Note that only half of the coefficients are useful in a real band-limited system because the remaining coefficients represent frequencies above one half of the sampling rate.

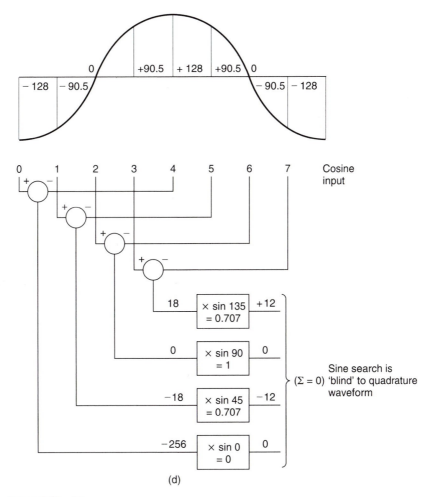

Figure 3.21 (d)

In STFTs the overlapping input sample blocks must be multiplied by window functions. The principle is the same as for the application in FIR filters shown in section 3.1. Figure 3.23 shows that multiplying the search frequency by the window has exactly the same result except that this need be done only once and much computation is saved. Thus in the STFT the basis function is a windowed sine or cosine wave.

The FFT is used extensively in such applications as phase correlation, where the accuracy with which the phase of signal components can be analysed is essential. It also forms the foundation of the discrete cosine transform.

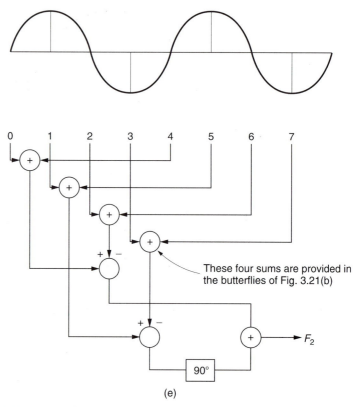

Figure 3.21 (e)

3.7   The discrete cosine transform (DCT)

The DCT is a special case of a discrete Fourier transform in which the sine components of the coefficients have been eliminated leaving a single number. This is actually quite easy. Figure 3.24(a) shows a block of input samples to a transform process. By repeating the samples in a time-reversed order and performing a discrete Fourier transform on the double-length sample set a DCT is obtained. The effect of mirroring the input waveform is to turn it into an even function whose sine coefficients are all zero. The result can be understood by considering the effect of individually transforming the input block and the reversed block.

Figure 3.24(b) shows that the phase of all the components of one block are in the opposite sense to those in the other. This means that when the components are added to give the transform of the double length block all the sine components cancel out, leaving only the cosine coefficients, hence the name of the transform.[9] In practice the sine component calculation is eliminated. Another advantage is that doubling the block length by mirroring doubles the frequency resolution, so that twice as

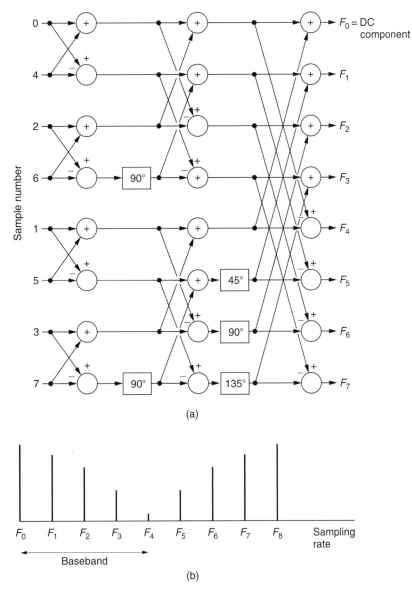

(a)

(b)

Figure 3.22 In (a) is the full butterfly diagram for an FFT. The spectrum this computes is shown in (b).

many useful coefficients are produced. In fact a DCT produces as many useful coefficients as input samples.

For image processing two-dimensional transforms are needed. In this case for every horizontal frequency, a search is made for all possible vertical frequencies. A two-dimensional DCT is shown in Figure 3.25. The DCT is separable in that the two-dimensional DCT can be obtained by computing in each dimension separately. Fast DCT algorithms are available.[10]

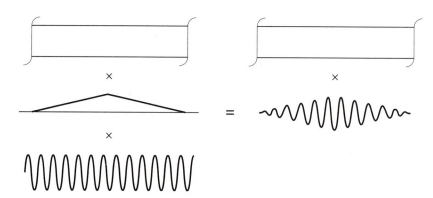

Figure 3.23 Multiplication of a windowed block by a sine wave basis function is the same as multiplying the raw data by a windowed basis function but requires less multiplication as the basis function is constant and can be pre-computed.

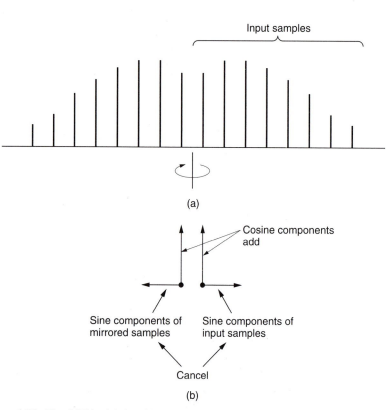

Figure 3.24 The DCT is obtained by mirroring the input block as shown in (a) prior to an FFT. The mirroring cancels out the sine components as in (b), leaving only cosine coefficients.

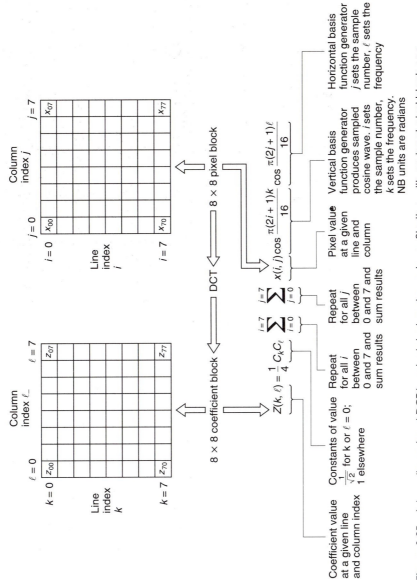

Figure 3.25 A two-dimensional DCT is calculated as shown here. Starting with an input pixel block one calculation is necessary to find a value for each coefficient. After 64 calculations using different basis functions the coefficient block is complete.

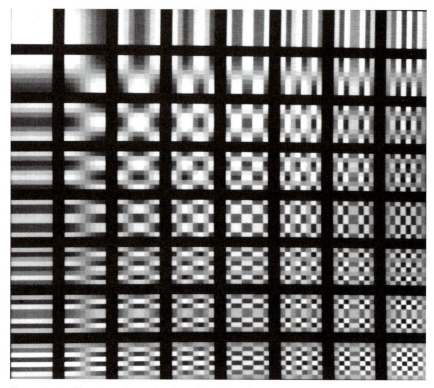

Figure 3.26 The discrete cosine transform breaks up an image area into discrete frequencies in two dimensions. The lowest frequency can be seen here at the top-left corner. Horizontal frequency increases to the right and vertical frequency increases downwards.

Figure 3.26 shows how a two-dimensional DCT is calculated by multiplying each pixel in the input block by terms which represent sampled cosine waves of various spatial frequencies. A given DCT coefficient is obtained when the result of multiplying every input pixel in the block is summed. Although most compression systems, including JPEG and MPEG, use square DCT blocks, this is not a necessity and rectangular DCT blocks are possible and are used in, for example, Digital Betacam, SX and DVC.

The DCT is primarily used in MPEG-2 because it converts the input waveform into a form where redundancy can be easily detected and removed. More details of the DCT can be found in Chapter 5.

3.8 The wavelet transform

The wavelet transform was not discovered by any one individual, but has evolved via a number of similar ideas and was only given a strong mathematical foundation relatively recently.[11,12] The wavelet transform is similar to the Fourier transform in that it has basis functions of various frequencies which are multiplied by the input waveform to identify the frequencies it contains. However, the Fourier transform is based on periodic signals and endless basis functions and requires windowing. The wavelet transform is fundamentally windowed, as the basis functions employed are not endless sine waves, but are finite on the time axis; hence the name. Wavelet transforms do not use a fixed window, but instead the window period is inversely proportional to the frequency being analysed. As a result a useful combination of time and frequency resolutions is obtained. High frequencies corresponding to transients in audio or edges in video are transformed with short basis functions and therefore are accurately located. Low frequencies are transformed with long basis functions which have good frequency resolution.

Fourier transform   Wavelet transform

Figure 3.27 Unlike discrete Fourier transforms, wavelet basis functions are scaled so that they contain the same number of cycles irrespective of frequency. As a result their frequency discrimination ability is a constant proportion of the centre frequency.

Figure 3.27 shows that that a set of wavelets or basis functions can be obtained simply by scaling (stretching or shrinking) a single wavelet on the time axis. Each wavelet contains the same number of cycles such that as the frequency reduces the wavelet gets longer. Thus the frequency discrimination of the wavelet transform is a constant fraction of the signal frequency. In a filter bank such a characteristic would be described as 'constant Q'. Figure 3.28 shows the division of the frequency domain by a wavelet transform is logarithmic whereas in the Fourier transform the division is uniform. The logarithmic coverage is effectively dividing the frequency domain into octaves and as such parallels the frequency discrimination of human hearing. For a comprehensive introduction to wavelets having exhaustive references, the reader is referred to Rioul and Vetterli.[13]

As it is relatively recent, the wavelet transform has yet to be widely used although it shows great promise. It has been successfully used in audio and in commercially available non-linear video editors and in other fields such as radiology and geology.

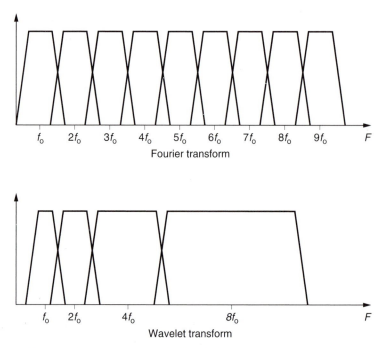

Figure 3.28 Wavelet transforms divide the frequency domain into octaves instead of the equal bands of the Fourier transform.

In video, wavelet compression does not display the 'blocking' of DCT-based coders at high compression factors; instead compression error is spread over the spectrum and appears as white noise.[14] It is naturally a multi-resolution transform allowing scaleable decoding

3.9   Motion compensation

Video data reduction relies on removing redundancy from the source signal. While redundancy can be removed from individual images, it will be seen that higher compression factors are only achieved by eliminating data which are common in successive images, so that only data which are different need to be sent. Motion causes the image to move with respect to the sampling grid, causing all the sample values in a moving area to change and preventing effective reduction. With motion estimation the image movement can be cancelled because the comparison can be made along the motion axis rather than the time axis. Greater reduction factors are then possible because it is then only necessary to send the motion parameters and a small number of genuine image differences.

Noise reduction in video signals works by combining together successive frames on the time axis such that the image content of the signal reinforces strongly whereas the random element in the signal due to noise does not. The noise reduction increases with the number of frames over which the noise is integrated, but image motion prevents simple combining of frames. If motion estimation is available, the image of a moving object in a particular frame can be integrated from the images in several frames which have been superimposed on the same part of the screen by displacements derived from the motion measurement. The result is that greater reduction of noise becomes possible.

3.10   Motion-estimation techniques

There are three main methods of motion estimation which are to be found in various applications: block matching, gradient matching and phase correlation. Each have their own characteristics which are quite different.

3.10.1   Block matching

This is the simplest technique to follow. In a given picture, a block of pixels is selected and stored as a reference. If the selected block is part of a moving object, a similar block of pixels will exist in the next picture, but not in the same place. As Figure 3.29 shows, block matching simply moves the reference block around over the second picture looking for matching pixel values. When a match is found, the displacement needed to obtain it is used as a basis for a motion vector.

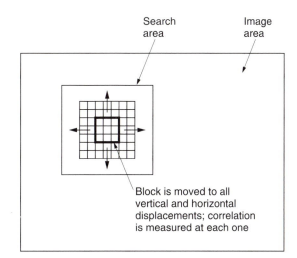

Figure 3.29 In block matching the search block has to be positioned at all possible relative motions within the search area and a correlation measured at each one.

While simple in concept, block matching requires an enormous amount of computation because every possible motion must be tested over the assumed range. Thus if the object is assumed to have moved over a 16-pixel range, then it will be necessary to test 16 different horizontal displacements in each of sixteen vertical positions; in excess of 65 000 positions. At each position every pixel in the block must be compared with every pixel in the second picture. In typical video displacements of twice the figure quoted here may be found, particularly in sporting events, and the computation then required becomes enormous.

One way of reducing the amount of computation is to perform the matching in stages where the first stage is inaccurate but covers a large motion range but the last stage is accurate but covers a small range.[15] The first matching stage is performed on a heavily filtered and subsampled picture, which contains far fewer pixels. When a match is found, the displacement is used as a basis for a second stage which is performed with a less heavily filtered picture. Eventually the last stage takes place to any desired accuracy. This hierarchical approach does reduce the computation required, but it suffers from the problem that the filtering of the first stage may make small objects disappear and they can never be found by subsequent stages if they are moving with respect to their background. Many televised sports events contain small, fast-moving objects. As the matching process depends upon finding similar luminance values, this can be confused by objects moving into shade or fades.

The simple block matching systems described above can only measure motion to the nearest pixel. If more accuracy is required, interpolators will be needed to shift the image by sub-pixel distances before attempting a match. The complexity rises once more. In compression systems an accuracy of half a pixel is accurate enough for most purposes.

3.10.2   Gradient matching

At some point in a picture, the function of brightness with respect to distance across the screen will have a certain slope, known as the spatial luminance gradient. If the associated picture area is moving, the slope will traverse a fixed point on the screen and the result will be that the brightness now changes with respect to time. This is a temporal luminance gradient. Figure 3.30 shows the principle. For a given spatial gradient, the temporal gradient becomes steeper as the speed of movement increases. Thus motion speed can be estimated from the ratio of the spatial and temporal gradients.[16]

The method only works well if the gradient remains essentially constant over the displacement distance; a characteristic which is not necessarily present in real video. In practice there are numerous processes which can change the luminance gradient. When an object moves so as to

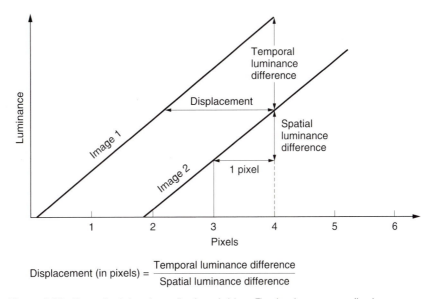

$$\text{Displacement (in pixels)} = \frac{\text{Temporal luminance difference}}{\text{Spatial luminance difference}}$$

Figure 3.30 The principle of gradient matching. The luminance gradient across the screen is compared with that through time.

obscure or reveal the background, the spatial gradient will change from field to field even if the motion is constant. Variations in illumination, such as when an object moves into shade, also cause difficulty.

The process can be assisted by recursion, in which the motion is estimated over a larger number of fields, but this will result in problems at cuts.

3.10.3 Phase correlation

Phase correlation works by performing a discrete Fourier transform on two successive fields and then subtracting all of the phases of the spectral components. The phase differences are then subject to a reverse transform which directly reveals peaks whose positions correspond to motions between the fields.[15,17] The nature of the transform domain means that if the distance and direction of the motion is measured accurately, the area of the screen in which it took place is not. Thus in practical systems the phase correlation stage is followed by a matching stage not dissimilar to the block matching process. However, the matching process is steered by the motions from the phase correlation, and so there is no need to attempt to match at all possible motions. By attempting matching on measured motion the overall process is made much more efficient.

One way of considering phase correlation is that by using the Fourier transform to break the picture into its constituent spatial frequencies. The hierarchical structure of block matching at various resolutions is in fact performed in parallel. In this way small objects are not missed because they will generate high-frequency components in the transform.

Although the matching process is simplified by adopting phase correlation, the Fourier transforms themselves require complex calculations. The high performance of phase correlation would remain academic if it were too complex to put into practice. However, if realistic values are used for the motion speeds which can be handled, the computation required by block matching actually exceeds that required for phase correlation.

The elimination of amplitude information from the phase correlation process ensures that motion estimation continues to work in the case of fades, objects moving into shade or flashguns firing.

The details of the Fourier transform have been described in section 3.5. A one-dimensional example of phase correlation will be given here by way of introduction. A line of luminance, which in the digital domain consists of a series of samples, is a function of brightness with respect to distance across the screen. The Fourier transform converts this function into a spectrum of spatial frequencies (units of cycles per picture width) and phases.

All television signals must be handled in linear-phase systems. A linear-phase system is one in which the delay experienced is the same for all frequencies. If video signals pass through a device which does not exhibit linear phase, the various frequency components of edges become displaced across the screen. Figure 3.31 shows what phase linearity means. If the left-hand end of the frequency axis (DC) is considered to be firmly anchored, but the right-hand end can be rotated to represent a change of position across the screen, it will be seen that as the axis twists evenly the result is phase shift proportional to frequency. A system having this characteristic is said to display linear phase.

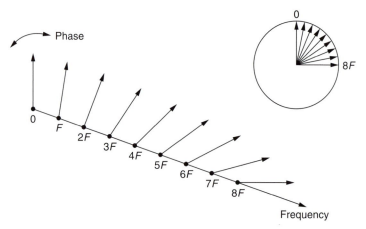

Figure 3.31   The definition of phase linearity is that phase shift is proportional to frequency. In phase-linear systems the waveform is preserved, and simply moves in time or space.

In the spatial domain, a phase shift corresponds to a physical movement. Figure 3.32 shows that if between fields a waveform moves along the line, the lowest frequency in the Fourier transform will suffer a given phase shift, twice that frequency will suffer twice that phase shift and so on. Thus it is potentially possible to measure movement between two successive fields if the phase differences between the Fourier spectra are analysed. This is the basis of phase correlation.

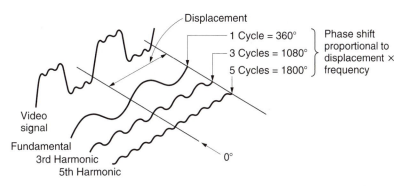

Figure 3.32 In a phase-linear system, shifting the video waveform across the screen causes phase shifts in each component proportional to frequency.

Figure 3.33 shows how a one-dimensional phase correlator works. The Fourier transforms of two lines from successive fields are computed and expressed in polar (amplitude and phase) notation (see section 3.5). The phases of one transform are all subtracted from the phases of the same frequencies in the other transform. Any frequency component having significant amplitude is then normalized, or boosted to full amplitude.

The result is a set of frequency components which all have the same amplitude, but have phases corresponding to the difference between two fields. These coefficients form the input to an inverse transform. Figure 3.34(a) shows what happens. If the two fields are the same, there are no phase differences between the two, and so all the frequency components

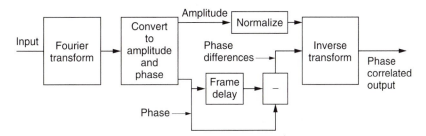

Figure 3.33 The basic components of a phase correlator.

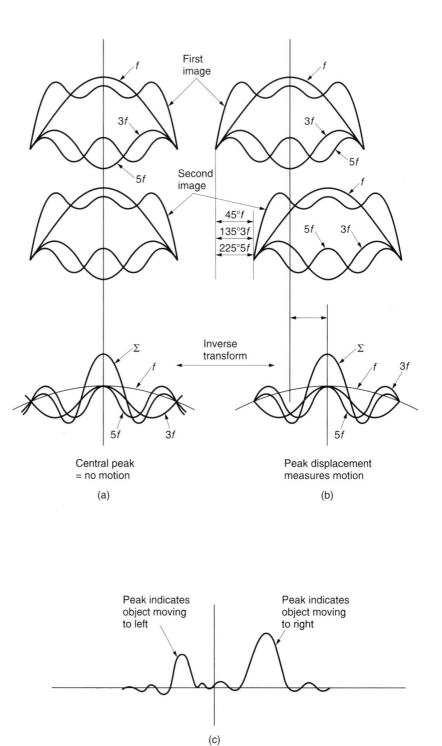

Figure 3.34   (a) The peak in the inverse transform is central for no motion. (b) In the case of motion the peak shifts by the distance moved. (c) If there are several motions, each one results in a peak.

are added with 0° phase to produce a single peak in the centre of the inverse transform. If, however, there was motion between the two fields, such as a pan, all the components will have phase differences, and this results in a peak shown in Figure 3.34(b) which is displaced from the centre of the inverse transform by the distance moved. Phase correlation thus actually measures the movement between fields.

In the case where the line of video in question intersects objects moving at different speeds, Figure 3.34(c) shows that the inverse transform would contain one peak corresponding to the distance moved by each object.

While this explanation has used one dimension for simplicity, in practice the entire process is two-dimensional. A two-dimensional Fourier transform of each field is computed, the phases are subtracted, and an inverse two-dimensional transform is computed, the output of which is a flat plane out of which three-dimensional peaks rise. This is known as a correlation surface.

Figure 3.35(a) shows some examples of a correlation surface. In (a) there has been no motion between fields and so there is a single central

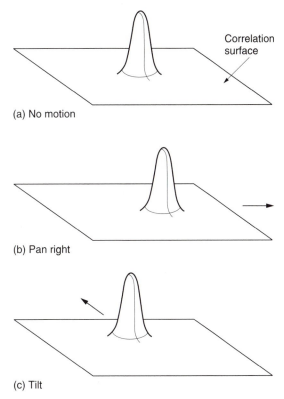

(a) No motion

(b) Pan right

(c) Tilt

Figure 3.35   (a) A two-dimensional correlation surface has a central peak when there is no motion. (b) In the case of a pan, the peak moves laterally. (c) A camera tilt moves the peak at right angles to the pan.

peak. In (b) there has been a pan and the peak moves across the surface. In (c) the camera has been depressed and the peak moves upwards.

Where more complex motions are involved, perhaps with several objects moving in different directions and/or at different speeds, one peak will appear in the correlation surface for each object.

It is a fundamental strength of phase correlation that it actually measures the direction and speed of moving objects rather than estimating, extrapolating or searching for them. The motion can be measured to sub-pixel accuracy without excessive complexity.

However, the consequences of uncertainty are that accuracy in the transform domain is incompatible with accuracy in the spatial domain. Although phase correlation accurately measures motion speeds and directions, it cannot specify where in the picture these motions are taking place. It is necessary to look for them in a further matching process. The efficiency of this process is dramatically improved by the inputs from the phase correlation stage.

3.11   Compression and requantizing

In compression systems the goal is to achieve coding gain by using fewer bits to represent the same information. The bandsplitting filters and transform techniques decribed earlier in this chapter *do not* achieve any coding gain. Their job is to express the information in a form in which redundancy can be identified. Paradoxically the output of a transform or a filter actually has a longer wordlength than the input because the integer input samples are multiplied by fractions. These processes have actually increased the redundancy in the signal. This section is concerned with the subsequent stage where the compression actually takes place.

Coding gain is obtained in one simple way: by shortening the wordlength of data words so that fewer bits are needed. These data words may be waveform samples in a sub-band-based system or coefficients in a transform-based system. In both cases positive and negative values must be handled. Figure 3.36(a) shows various signal levels in two's complement coding. As the level falls a phenomenon called *sign extension* takes place where more and more bits at the most significant end of the word simply copy the sign bit (which is the MSB). Coding gain can be obtained by eliminating the redundant sign extension bits as Figure 3.36(b) shows.

Taking bits out of the middle of a word is not straightforward and in practice the solution is to multiply by a level-dependent factor to eliminate the sign extension bits. If this is a power of 2 the useful bits will simply shift left up to the sign bit. The right-hand zeros are simply omitted from the transmission. On decoding a compensating division

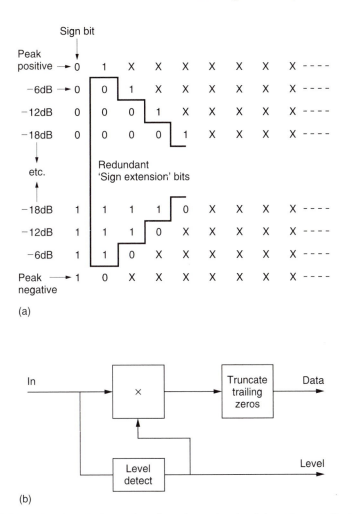

Figure 3.36 In two's complement coding, sign extension takes place as the level falls. These sign-extended bits are redundant and can be eliminated by multiplying by a level-dependent factor and neglecting the trailing zeros as shown in (b).

must be performed. The multiplication factor must be transmitted along with the compressed data so this can be done. Clearly if only the sign extension bits are eliminated this process is lossless because exactly the same data values are available at the decoder.

The reason for sub-band filtering and transform coding now becomes clear because in real signals the levels in most sub-bands and the value of most coefficients is considerably less than the highest level.

In many cases the coding gain obtained in this way will not be enough and the wordlength has to be shortened even more. Following the multiplication, a larger number of bits are rounded off from the least significant end of the word. The result is that the same signal range is retained but it is expressed with less accuracy. It is as if the original analog signal had been converted using fewer quantizing steps, hence the term requantizing.

During the decoding process an *inverse quantizer* will be employed to convert the compressed value back to its original form. Figure 3.37 shows a requantizer and an inverse quantizer. The inverse quantizer must divide by the same gain factor as was used in the compressor and re-insert trailing zeros up to the required wordlength.

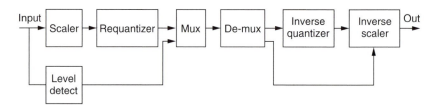

Figure 3.37 Coding gain is obtained by shortening the sample or coefficient wordlength so fewer bits are needed. The input values are scaled or amplified to near maximum amplitude prior to rounding off the low-order bits. The scale factor must be transmitted to allow the process to be reversed at the decoder.

A non-uniform quantization process may be used in which the quantizing steps become larger as the signal amplitude increases. As the quantizing steps are made larger, more noise will be suffered, but the noise is arranged to be generated at frequencies where it will not be perceived, or on signal values which occur relatively infrequently.

Shortening the wordlength of a sample from the LSB end reduces the number of quantizing intervals available without changing the signal amplitude. As Figure 3.38 shows, the quantizing intervals become larger and the original signal is *requantized* with the new interval structure. It will be seen that truncation does not meet the above requirement as it results in signal-dependent offsets because it always rounds in the same direction. Proper numerical rounding is essential for accuracy. Rounding in two's complement is a little more complex than in pure binary.

If the parameter to be requantized is a transform coefficient the result after decoding is that the frequency which is reproduced has an incorrect amplitude due to quantizing error. If all the coefficients describing a signal are requantized to roughly the same degree then the error will be uniformly present at all frequencies and will be noise-like.

However, if the data are time-domain samples, as in, for example, an audio sub-band coder, the result will be different. Although the original audio conversion would have been correctly dithered, the linearizing random element in the low-order bits will be some way below the end of

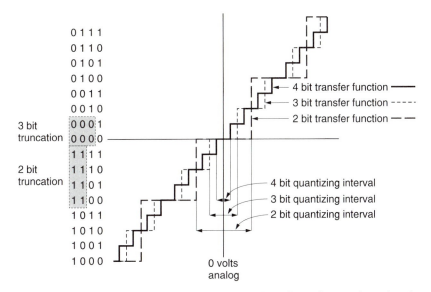

Figure 3.38 Shortening the wordlength of a sample reduces the number of codes which can describe the voltage of the waveform. This makes the quantizing steps bigger, hence the term requantizing. It can be seen that simple truncation or omission of low-order bits does not give analogous behaviour. Rounding is necessary to give the same result as if the larger steps had been used in the original conversion.

the shortened word. If the word is simply rounded to the nearest integer the linearizing effect of the original dither will be lost and the result will be quantizing distortion. As the distortion takes place in a bandlimited system the harmonics generated will alias back within the band. Where the requantizing process takes place in a sub-band, the distortion products will be confined to that sub-band as shown in Figure 3.39.

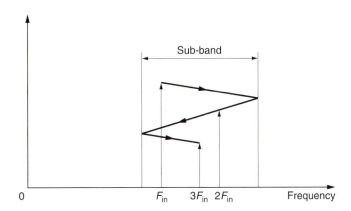

Figure 3.39 Requantizing a band-limited signal causes harmonics which will always alias back within the band.

In practice, the wordlength of samples should be shortened in such a way that the requantizing error is converted to noise rather than distortion. One technique which meets this requirement is to use digital dithering[18] prior to rounding.

Digital dither is a pseudo-random sequence of numbers. If it is required to simulate the analog dither signal of Figure 2.25, then it is obvious that the noise must be bipolar so that it can have an average voltage of zero. Two's complement coding must be used for the dither values.

Figure 3.40 shows a simple digital dithering system for shortening sample wordlength. The output of a two's complement pseudo-random sequence generator of appropriate wordlength is added to input samples

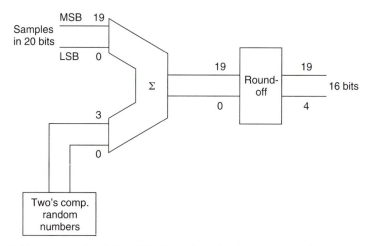

Figure 3.40 In a simple digital dithering system, two's complement values from a random number generator are added to low-order hits of the input. The dithered values are then rounded up or down according to the value of the bits to be removed. The dither linearizes the requantizing.

prior to rounding. The most significant of the bits to be discarded is examined in order to determine whether the bits to be removed sum to more or less than half a quantizing interval. The dithered sample is either rounded down, i.e. the unwanted bits are simply discarded, or rounded up, i.e. the unwanted bits are discarded but one is added to the value of the new short word. The rounding process is no longer deterministic because of the added dither which provides a linearizing random component.

The probability density of the pseudo-random sequence is important. Lipshitz *et al.*[19] found that uniform probability density produced noise modulation, in which the amplitude of the random component varies as a function of the amplitude of the samples. A triangular probability density function obtained by adding together two pseudo-random sequences eliminated the noise modulation to yield a signal-independent white-noise component in the least significant bit.

References

1. van den Enden, A.W.M. and Verhoeckx, N.A.M., Digital signal processing: theoretical background. *Philips Tech. Rev.*, **42**, 110–144, (1985)
2. McClellan, J.H., Parks, T.W. and Rabiner, L.R., A computer program for designing optimum FIR linear-phase digital filters. *IEEE Trans. Audio and Electroacoustics*, **AU-21**, 506–526 (1973)
3. Jayant, N.S. and Noll, P., *Digital Coding of Waveforms: Principles and Applications to Speech and Video*. Englewood Cliffs, NJ: Prentice Hall (1984)
4. Theile, G., Stoll, G. and Link, M., Low bit rate coding of high quality audio signals: an introduction to the MASCAM system. *EBU Tech. Review*, No. 230, 158–181 (1988)
5. Chu, P.L., Quadrature mirror filter design for an arbitrary number of equal bandwidth channels. *IEEE Trans. ASSP*, **ASSP-33**, 203–218 (1985)
6. Fettweis, A., Wave digital filters: Theory and practice. *Proc. IEEE*, **74**, 270–327 (1986)
7. Weiss, P. and Christensson, J., Real time implementation of sub-pixel motion estimation for broadcast applications. *IEE Digest*, 1990/128
8. Kraniauskas, P., *Transforms in Signals and Systems*, Chapter 6, Wokingham: Addison-Wesley (1992)
9. Ahmed, N., Natarajan, T. and Rao, K., Discrete cosine transform. *IEEE Trans. Computers*, **C-23** 90–93 (1974)
10. De With, P.H.N., Data compression techniques for digital video recording. PhD thesis, Technical University of Delft (1992)
11. Goupillaud, P., Grossman, A. and Morlet, J., Cycle-octave and related transforms in seismic signal analysis. *Geoexploration*, **23**, 85–102, Elsevier (1984/5)
12. Daubechies, I., The wavelet transform, time–frequency localisation and signal analysis. *IEEE Trans. Info. Theory*, **36**, No.5, 961–1005 (1990)
13. Rioul, O. and Vetterli, M., Wavelets and signal processing. *IEEE Signal Process. Mag.*, 14–38 (Oct. 1991)
14. Huffman, J., Wavelets and image compression. Presented at 135th SMPTE Tech. Conf. (Los Angeles 1993) Preprint No. 135–98
15. Thomas, G.A., Television motion measurement for DATV and other applications. *BBC Res. Dept. Rept*, RD 1987/11 (1987)
16. Limb, J.O. and Murphy, J.A., Measuring the speed of moving objects from television signals. *IEEE Trans. Commun*, 474–478 (1975)
17. Pearson, J.J. *et al.*, Video rate image correlation processor. *S.P.I.E.*, Vol.119, *Application of digital image processing*, IOCC (1977)
18. Vanderkooy, J. and Lipshitz, S.P., Digital dither. Presented at the 81st Audio Engineering Society Convention (Los Angeles, 1986), Preprint 2412(C-8)
19. Lipshitz, S.P., Wannamaker, R.A. and Vanderkooy, J., Quantization and dither: a theoretical survey. *J. Audio Eng. Soc.*, **40**, 355–375 (1992)

4

Audio compression

4.1 Introduction

Physics can tell us the mechanism by which disturbances propagate through the air. If this is our definition of sound, we have the problem that in physics there are no limits to the frequencies and levels which must be considered. Biology can tell that the ear only responds to a certain range of frequencies provided a threshold level is exceeded. This is a better definition of sound; reproduction is easier because it is only necessary to reproduce that range of levels and frequencies which the ear can detect.

Psychoacoustics can describe how our hearing has finite resolution in both time and frequency domains such that what we perceive is an inexact impression. Some aspects of the original disturbance are inaudible to us and are said to be masked. If our goal is the highest quality, we can design our imperfect equipment so that the shortcomings are masked. Conversely if our goal is economy we can use compression and hope that masking will disguise the inaccuracies it causes.

By definition, the sound quality of a perceptive coder can only be assessed by human hearing. Equally, a useful perceptive coder can only be designed with a good knowledge of the human hearing mechanism.[1] The acuity of the human ear is astonishing. The frequency range is extremely wide, covering some ten octaves (an octave is a doubling of pitch or frequency) without interruption. It can detect tiny amounts of distortion, and will accept an enormous dynamic range. If the ear detects a different degree of impairment between two codecs having the same bit rate in properly conducted tests, we can say that one of them is superior. Thus quality is completely subjective and can only be checked by listening tests. However, any characteristic of a signal which can be heard can also be measured by a suitable instrument. The subjective tests can tell us how sensitive the instrument should be. Then the objective readings from the instrument give an indication of how acceptable a signal is in respect of that characteristic. Instruments for assessing the performance of codecs are currently extremely rare and there remains much work to be done.

4.2 The ear

The sense we call hearing results from acoustic, mechanical, hydraulic, nervous and mental processes in the ear/brain combination, leading to the term psychoacoustics. It is only possible briefly to introduce the subject here. The interested reader is referred to Moore[2] for an excellent treatment.

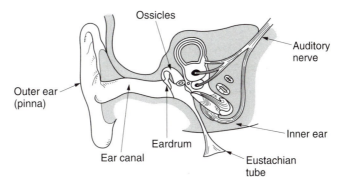

Figure 4.1 The structure of the human ear. See text for details.

Figure 4.1 shows that the structure of the ear is traditionally divided into the outer, middle and inner ears. The outer ear works at low impedance, the inner ear works at high impedance, and the middle ear is an impedance matching device. The visible part of the outer ear is called the pinna which plays a subtle role in determining the direction of arrival of sound at high frequencies. It is too small to have any effect at low frequencies. Incident sound enters the auditory canal or meatus. The pipe-like meatus causes a small resonance at around 4 KHz. Sound vibrates the eardrum or tympanic membrane which seals the outer ear from the middle ear. The inner ear or cochlea works by sound travelling though a fluid. Sound enters the cochlea via a membrane called the oval window. If airborne sound were to be incident on the oval window directly, the serious impedance mismatch would cause most of the sound to be reflected. The middle ear remedies that mismatch by providing a mechanical advantage. The tympanic membrane is linked to the oval window by three bones known as ossicles which act as a lever system such that a large displacement of the tympanic membrane results in a smaller displacement of the oval window but with greater force. Figure 4.2 shows that the malleus applies a tension to the tympanic membrane rendering it conical in shape. The malleus and the incus are firmly joined together to form a lever. The incus acts upon the stapes through a spherical joint. As the area of the tympanic membrane is greater than that of the oval window, there is a further multiplication of the available force. Consequently small pressures over the large area of the tympanic membrane are converted to high pressures over the small area of the oval window.

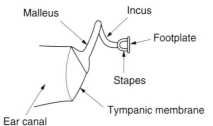

Figure 4.2   The malleus tensions the tympanic membrane into a conical shape. The ossicles provide an impedance-transforming lever system between the tympanic membrane and the oval.

The middle ear is normally sealed, but ambient pressure changes will cause static pressure on the tympanic membrane which is painful. The pressure is relieved by the Eustachian tube which opens involuntarily while swallowing. The Eustachian tubes open into the cavities of the head and must normally be closed to avoid one's own speech appearing deafeningly loud.

The ossicles are located by minute muscles which are normally relaxed. However, the middle ear reflex is an involuntary tightening of the *tensor tympani* and *stapedius* muscles which heavily damp the ability of the tympanic membrane and the stapes to transmit sound by about 12 dB at frequencies below 1 KHz. The main function of this reflex is to reduce the audibility of one's own speech. However, loud sounds will also trigger this reflex which takes some 60–120 milliseconds to occur, too late to protect against transients such as gunfire.

4.3   The cochlea

The cochlea, shown in Figure 4.3(a), is a tapering spiral cavity within bony walls which is filled with fluid. The widest part, near the oval window, is called the *base* and the distant end is the *apex*. Figure 4.3(b) shows that the cochlea is divided lengthwise into three volumes by Reissner's membrane and the basilar membrane. The *scala vestibuli* and the *scala tympani* are connected by a small aperture at the apex of the cochlea known as the *helicotrema*. Vibrations from the stapes are transferred to the oval window and become fluid pressure variations which are relieved by the flexing of the round window. Effectively the basilar membrane is in series with the fluid motion and is driven by it except at very low frequencies where the fluid flows through the helicotrema, bypassing the basilar membrane.

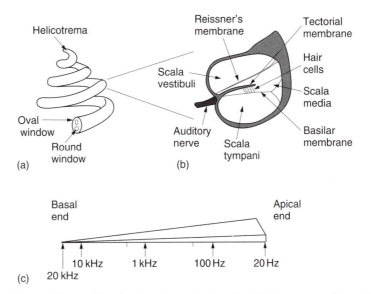

Figure 4.3 (a) The cochlea is a tapering spiral cavity. (b) The cross-section of the cavity is divided by Reissner's membrane and the basilar membrane. (c) The basilar membrane tapers so its resonant frequency changes along its length.

Figure 4.3(c) shows that the basilar membrane is not uniform, but tapers in width and varies in thickness in the opposite sense to the taper of the cochlea. The part of the basilar membrane which resonates as a result of an applied sound is a function of the frequency. High frequencies cause resonance near to the oval window, whereas low frequencies cause resonances further away. More precisely the distance from the apex where the maximum resonance occurs is a logarithmic function of the frequency. Consequently tones spaced apart in octave steps will excite evenly spaced resonances in the basilar membrane. The prediction of resonance at a particular location on the membrane is called *place theory*. Essentially the basilar membrane is a mechanical frequency analyser. A knowledge of the way it operates is essential to an understanding of musical phenomena such as pitch discrimination, timbre, consonance and dissonance and to auditory phenomena such as critical bands, masking and the precedence effect.

The vibration of the basilar membrane is sensed by the organ of Corti which runs along the centre of the cochlea. The organ of Corti is active in that it contains elements which can generate vibration as well as sense it. These are connected in a regenerative fashion so that the Q factor, or frequency selectivity, of the ear is higher than it would otherwise be. The deflection of hair cells in the organ of Corti triggers nerve firings and these signals are conducted to the brain by the auditory nerve.

Nerve firings are not a perfect analog of the basilar membrane motion. A nerve firing appears to occur at a constant phase relationship to the basilar vibration; a phenomenon called phase locking, but firings do not necessarily occur on every cycle. At higher frequencies firings are intermittent, yet each is in the same phase relationship.

The resonant behaviour of the basilar membrane is not observed at the lowest audible frequencies below 50 Hz. The pattern of vibration does not appear to change with frequency and it is possible that the frequency is low enough to be measured directly from the rate of nerve firings.

4.4   Level and loudness

At its best, the ear can detect a sound pressure variation of only 2×10^{-5} Pascals r.m.s. and so this figure is used as the reference against which sound pressure level (SPL) is measured. The sensation of loudness is a logarithmic function of SPL and consequently a logarithmic unit, the deciBel, is used in audio measurement.

The dynamic range of the ear exceeds 130 dB, but at the extremes of this range, the ear is either straining to hear or is in pain. Neither of these cases can be described as pleasurable or entertaining, and it is hardly necessary to produce audio of this dynamic range since, among other things, the consumer is unlikely to have anywhere sufficiently quiet to listen to it. On the other hand, extended listening to music whose dynamic range has been excessively compressed is fatiguing.

The frequency response of the ear is not at all uniform and it also changes with SPL. The subjective response to level is called loudness and is measured in *phons*. The phon scale and the SPL scale coincide at 1 KHz, but at other frequencies the phon scale deviates because it displays the actual SPLs judged by a human subject to be equally loud as a given level at 1 KHz. Figure 4.4 shows the so-called equal loudness contours which were originally measured by Fletcher and Munson and subsequently by Robinson and Dadson. Note the irregularities caused by resonances in the meatus at about 4 KHz and 13 KHz.

Usually, people's ears are at their most sensitive between about 2 KHz and 5 KHz, and although some people can detect 20 KHz at high level, there is much evidence to suggest that most listeners cannot tell if the upper frequency limit of sound is 20 KHz or 16 KHz.[3,4] For a long time it was thought that frequencies below about 40 Hz were unimportant, but it is now clear that reproduction of frequencies down to 20 Hz improves reality and ambience.[5] The generally accepted frequency range for high-quality audio is 20 Hz to 20 000 Hz, although for broadcasting an upper limit of 15 000 Hz is often applied.

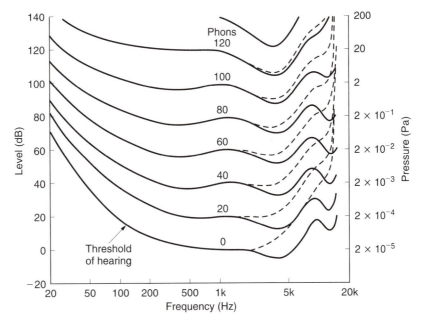

Figure 4.4 Contours of equal loudness showing that the frequency response of the ear is highly level dependent (solid line. age 20; dashed line, age 60).

The most dramatic effect of the curves of Figure 4.4 is that the bass content of reproduced sound is disproportionately reduced as the level is turned down.

Loudness is a subjective reaction and is almost impossible to measure. In addition to the level-dependent frequency response problem, the listener uses the sound not for its own sake but to draw some conclusion about the source. For example, most people hearing a distant motorcycle will describe it as being loud. Clearly at the source, it *is* loud, but the listener has compensated for the distance.

The best that can be done is to make some compensation for the level-dependent response using *weighting curves*. Ideally there should be many, but in practice the A, B and C weightings were chosen where the A curve is based on the 40-phon response. The measured level after such a filter is in units of dBA. The A curve is almost always used because it most nearly relates to the annoyance factor of distant noise sources.

4.5   Frequency discrimination

Figure 4.5 shows an uncoiled basilar membrane with the apex on the left so that the usual logarithmic frequency scale can be applied. The envelope of displacement of the basilar membrane is shown for a single frequency in (a). The vibration of the membrane in sympathy with a single frequency cannot be localized to an infinitely small area, and nearby areas are forced to vibrate at the same frequency with an amplitude that decreases with distance. Note that the envelope is asymmetrical because the membrane is tapering and because of frequency-dependent losses in the propagation of vibrational energy down the cochlea. If the frequency is changed, as in (b), the position of maximum displacement will also change. As the basilar membrane is continuous, the position of maximum displacement is infinitely variable allowing extremely good pitch discrimination of about one-twelfth of a semitone which is determined by the spacing of hair cells.

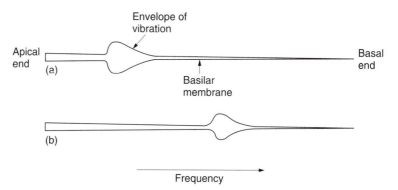

Figure 4.5   The basilar membrane symbolically uncoiled. (a) Single frequency causes the vibration envelope shown. (b) Changing the frequency moves the peak of the envelope.

In the presence of a complex spectrum, the finite width of the vibration envelope means that the ear fails to register energy in some bands when there is more energy in a nearby band. Within those areas, other frequencies are mechanically excluded because their amplitude is insufficient to dominate the local vibration of the membrane. Thus the Q factor of the membrane is responsible for the degree of auditory masking, defined as the decreased audibility of one sound in the presence of another.

4.6   Critical bands

The term used in psychoacoustics to describe the finite width of the vibration envelope is *critical bandwidth*. Critical bands were first described by Fletcher.[6] The envelope of basilar vibration is a complicated function. It is clear from the mechanism that the area of the membrane involved will increase as the sound level rises. Figure 4.6 shows the bandwidth as a function of level.

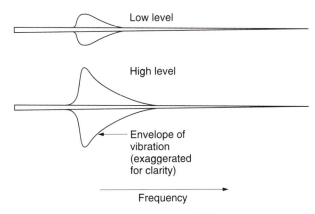

Figure 4.6   The critical bandwidth changes with SPL.

As was shown in Chapter 3, transform theory teaches that the higher the frequency resolution of a transform, the worse the time accuracy. As the basilar membrane has finite frequency resolution measured in the width of a critical band, it follows that it must have finite time resolution. This also follows from the fact that the membrane is resonant, taking time to start and stop vibrating in response to a stimulus. There are many examples of this. Figure 4.7 shows the impulse response. Figure 4.8 shows the perceived loudness of a tone burst increases with duration up to about 200 milliseconds due to the finite response time.

The ear has evolved to offer intelligibility in reverberant environments which it does by averaging all received energy over a period of about 30 milliseconds. Reflected sound which arrives within this time is integrated to produce a louder sensation, whereas reflected sound which arrives after that time can be temporally discriminated and is perceived as an echo. Our simple microphones have no such ability,

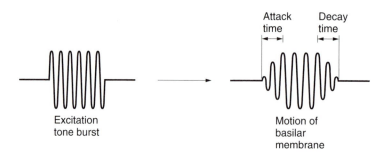

Figure 4.7 Impulse response of the ear showing slow attack and decay due to resonant behaviour.

which is why we often need to have acoustic treatment in areas where microphones are used.

A further example of the finite time discrimination of the ear is the fact that short interruptions to a continuous tone are difficult to detect. Finite time resolution means that masking can take place even when the masking tone begins after and ceases before the masked sound. This is referred to as forward and backward masking.[7]

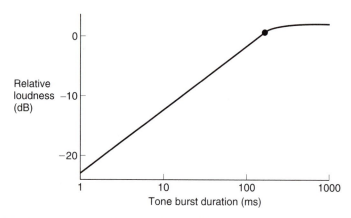

Figure 4.8 Perceived level of tone burst rises with duration as resonance builds up.

As the vibration envelope is such a complicated shape, Moore and Glasberg[8] have proposed the concept of equivalent rectangular bandwidth to simplify matters. The ERB is the bandwidth of a rectangular filter which passes the same power as a critical band. Figure 4.9(a) shows the expression they have derived linking the ERB with frequency. This is plotted in (b) where it will be seen that one third of an octave is a good approximation. This is about thirty times broader than the pitch discrimination also shown in (a).

Some treatments of human hearing liken the basilar membrane to a bank of fixed filters each of which has the width of a critical band. The

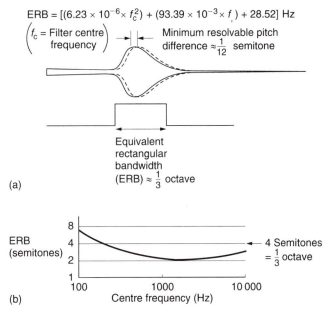

Figure 4.9 Effective rectangular bandwidth of critical band is much wider than the resolution of the pitch discrimination mechanism.

frequency response of such a filter can be deduced from the envelope of basilar displacement as has been done in Figure 4.10. The fact that no agreement has been reached on the number of such filters should alert the suspicions of the reader. The fact that a third octave filter bank model cannot explain pitch discrimination some thirty times better is another cause for doubt. The response of the basilar membrane is centred upon the input frequency and no fixed filter can do this. However, the most worrying aspect of the fixed filter model is that according to Figure 4.10(b) a single tone would cause a response in several bands which would be interpreted as several tones. This is at variance with reality. Far from masking higher frequencies, we appear to be creating them!

This author prefers to keep in mind how the basilar membrane is actually vibrating in response to an input spectrum. If a mathematical model of the ear is required, then it has to be described as performing a finite resolution continuous frequency transform.

4.7 Beats

Figure 4.11 shows an electrical signal (a) in which two equal sine waves of nearly the same frequency have been linearly added together. Note that the envelope of the signal varies as the two waves move in and out of phase. Clearly the frequency transform calculated to infinite accuracy is that shown in (b). The two amplitudes are constant and there is no

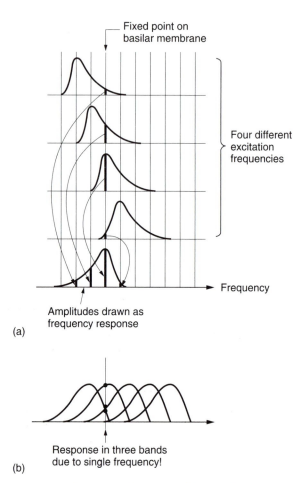

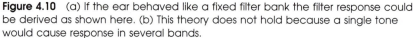

Figure 4.10 (a) If the ear behaved like a fixed filter bank the filter response could be derived as shown here. (b) This theory does not hold because a single tone would cause response in several bands.

evidence of the envelope modulation. However, such a measurement requires an infinite time. When a shorter time is available, the frequency discrimination of the transform falls and the bands in which energy is detected become broader. When the frequency discrimination is too wide to distinguish the two tones as in (c), the result is that they are registered as a single tone. The amplitude of the single tone will change from one measurement to the next because the envelope is being measured. The rate at which the envelope amplitude changes is called a *beat* frequency which is not actually present in the input signal. Beats are an artefact of finite frequency resolution transforms. The fact that human hearing produces beats from pairs of tones proves that it has finite resolution.

Measurement of when beats occur allows measurement of critical bandwidth. Figure 4.12 shows the results of human perception of a two-

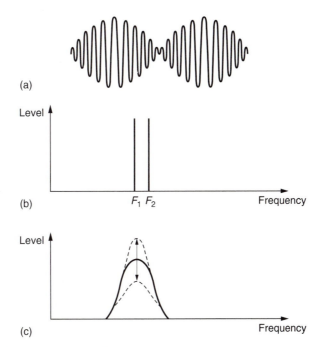

Figure 4.11 (a) Result of adding two sine waves of similar frequency. (b) Spectrum of (a) to infinite accuracy. (c) With finite accuracy only a single frequency is distinguished whose amplitude changes with the envelope of (a) giving rise to beats.

tone signal as the frequency **dF** difference changes. When **dF** is zero, described musically as *unison*, only a single note is heard. As **dF** increases, beats are heard, yet only a single note is perceived. The limited frequency resolution of the basilar membrane has *fused* the two tones together. As **dF** increases further, the sensation of beats ceases at 12–15 Hz and is replaced by a sensation of roughness or *dissonance*. The roughness is due to parts

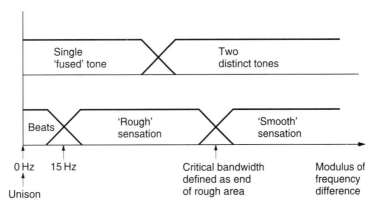

Figure 4.12 Perception of two-tone signal as frequency difference changes.

of the basilar membrane being unable to decide the frequency at which to vibrate. The regenerative effect may well become confused under such conditions. The roughness persists until **dF** has reached the critical bandwidth beyond which two separate tones will be heard because there are now two discrete basilar resonances. In fact this is the definition of critical bandwidth.

4.8   Codec level calibration

The functioning of the ear is noticeably level-dependent and perceptive coders take this into account. However, all signal processing takes place in the electrical domain with respect to electrical levels whereas the hearing mechanism operates with respect to sound pressure level. Figure 4.13 shows that in an ideal system the overall gain of the microphones and ADCs is such that the PCM codes have a relationship with sound pressure which is the same as that assumed by the model in the codec.

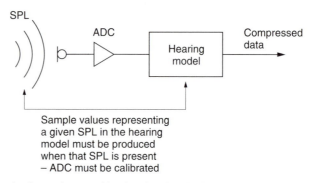

Figure 4.13   Audio coders must be level calibrated so that the psychoacoustic decisions in the coder are based on correct sound pressure levels.

Equally the overall gain of the DAC and loudspeaker system should be such that the sound pressure levels which the codec assumes are those actually heard. Clearly the gain control of the microphone and the volume control of the reproduction system must be calibrated if the hearing model is to function properly. If, for example, the microphone gain was too low and this was compensated by advancing the loudspeaker gain, the overall gain would be the same but the codec would be fooled into thinking that the sound pressure level was less than it really was and the masking model would not then be appropriate.

The above should come as no surprise as analog audio codecs such as the various Dolby systems have required and implemented line-up procedures and suitable tones.

4.9 Quality measurement

As was seen in Chapter 3, one way in which coding gain is obtained is to requantize sample values to reduce the wordlength. Since the resultant requantizing error results in energy moving from one frequency to another, the masking model is essential to estimate how audible the effect will be. The greater the degree of reduction required, the more precise the model must be. If the masking model is inaccurate, then equipment based upon it may produce audible artifacts under some circumstances. Artifacts may also result if the model is not properly implemented. As a result, development of data reduction units requires careful listening tests with a wide range of source material.[9,10] The presence of artifacts at a given compression factor indicates only that performance is below expectations; it does not distinguish between the implementation and the model. If the implementation is verified, then a more detailed model must be sought. Naturally comparative listening tests are only valid if all the codecs have been level calibrated.

Properly conducted listening tests are expensive and time consuming, and alternative methods have been developed which can be used objectively to evaluate the performance of different techniques. The noise-to-masking ratio (NMR) is one such measurement.[11] Figure 4.14

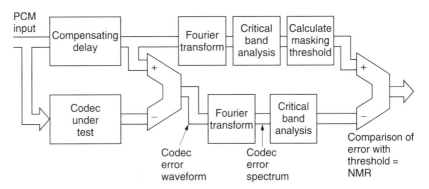

Figure 4.14 The noise-to-masking ratio is derived as shown here.

shows how NMR is measured. Input audio signals are fed simultaneously to a data reduction coder and decoder in tandem and to a compensating delay whose length must be adjusted to match the codec delay. At the output of the delay, the coding error is obtained by subtracting the codec output from the original. The original signal is spectrum analysed into critical bands in order to derive the masking threshold of the input audio, and this is compared with the critical band spectrum of the error. The NMR in each critical band is the ratio between the masking threshold and the quantizing error due to the codec. An average NMR for all bands can be computed. A positive NMR in any band indicates that artifacts are potentially audible. Plotting the average NMR against time is a powerful technique, as with an ideal codec the NMR should be stable with different types of program material. If this is not the case the codec could perform quite differently as a function of the source material. NMR excursions can be correlated with the waveform of the audio input to analyse how the extra noise was caused and to redesign the codec to eliminate it.

Practical systems should have a finite NMR in order to give a degree of protection against difficult signals which have not been anticipated and against the use of post-codec equalization or several tandem codecs which could change the masking threshold. There is a strong argument that professional devices should have a greater NMR than consumer or program delivery devices.

4.10   The limits

There are, of course, limits to all technologies. Eventually artifacts will be heard as the amount of compression is increased which no amount of detailed modelling will remove. The ear is only able to perceive a certain proportion of the information in a given sound. This could be called the perceptual entropy,[12] and all additional sound is redundant or irrelevant. Data reduction works by removing the redundancy, and clearly an ideal system would remove all of it, leaving only the entropy. Once this has been done, the masking capacity of the ear has been reached and the NMR has reached zero over the whole band. Reducing the data rate further must reduce the entropy, because raising noise further at any frequency will render it audible. In practice the audio bandwidth will have to be reduced in order to keep the noise level acceptable. In MPEG-1 pre-filtering allows data from higher sub-bands to be neglected. MPEG-2 has introduced some low sampling rate options for this purpose. Thus there is a limit to the degree of data reduction which can be achieved even with an ideal coder. Systems which go beyond that limit are not appropriate for high-quality music, but are relevant in news gathering and communications where intelligibility of speech is the criterion.

Interestingly, the data rate out of a coder is virtually independent of the input sampling rate unless the sampling rate is very low. This is because the entropy of the sound is in the waveform, not in the number of samples carrying it.

The compression factor of a coder is only part of the story. All codecs cause delay, and in general the greater the compression, the longer the delay. In some applications, such as telephony, a short delay is required.[13]

In many applications, the compressed channel will have a constant bit rate, and so a constant compression factor is required. In real program material, the entropy varies and so the NMR will fluctuate. If greater delay can be accepted, as in a recording application, memory buffering can be used to allow the coder to operate at constant NMR and instantaneously variable data rate. The memory absorbs the instantaneous data rate differences of the coder and allows a constant rate in the channel. A higher effective compression factor will then be obtained. Near-constant quality can also be achieved using statistical multiplexing.

4.11   Compression applications

One of the fundamental concepts of PCM audio is that the signal-to-noise ratio of the channel can be determined by selecting a suitable wordlength. In conjunction with the sampling rate, the resulting data rate is then determined. In many cases the full data rate can be transmitted or recorded. A high-quality digital audio channel requires around one megabit per second, whereas a standard definition component digital video channel needs two hundred times as much. With mild video compression the video bit rate is still so much in excess of the audio rate that audio compression is not worth while. Only when high video compression factors are used does it become necessary to apply compression to the audio.

Professional equipment traditionally has operated with higher sound quality than consumer equipment in order to allow for some inevitable quality loss in production, and so there will be less pressure to use data reduction. In fact there is even pressure against the use of audio compression in critical applications because of the loss of quality, particularly in stereo and surround sound.

With the rising density and falling cost of digital recording media, the use of compression in professional audio storage applications is hard to justify. However, where there is a practical or economic restriction on channel bandwidth compression becomes essential. Now that the technology exists to broadcast television pictures using compressed digital coding, it is obvious that the associated audio should be conveyed in the same way.

4.12   History of MPEG audio coding

The ISO (International Standards Organization) and the IEC (International Electrotechnical Commission) recognized that compression would have an important part to play in future digital video products and in 1988 established the ISO/IEC/MPEG (Moving Picture Experts Group) to compare and assess various coding schemes in order to arrive at an international standard. The terms of reference were extended the same year to include audio and the MPEG/Audio group was formed.

As part of the Eureka 147 project, a system known as MUSICAM[14] (Masking pattern adapted Universal Sub-band Integrated Coding and Multiplexing) was developed jointly by CCETT in France, IRT in Germany and Philips in the Netherlands. MUSICAM was designed to be suitable for DAB (digital audio broadcasting).

As a parallel development, the ASPEC[15] (Adaptive Spectral Perceptual Entropy Coding) system was developed from a number of earlier systems as a joint proposal by AT&T Bell Labs, Thomson, the Fraunhofer Society and CNET. ASPEC was designed for high degrees of compression to allow audio transmission on ISDN.

These two systems were both fully implemented by July 1990 when comprehensive subjective testing took place at the Swedish Broadcasting Corporation.[9,16,17] As a result of these tests, the MPEG/Audio group combined the attributes of both ASPEC and MUSICAM into a draft standard[18] having three levels of complexity and performance.

These three different levels are needed because of the number of possible applications. Audio coders can be operated at various compression factors with different quality expectations. Stereophonic classical music requires different quality criteria to monophonic speech. As was seen in Chapter 1, the complexity of the coder will be reduced with a smaller compression factor. For moderate compression, a simple codec will be more cost effective. On the other hand, as the compression factor is increased, it will be necessary to employ a more complex coder to maintain quality.

At each level, MPEG coding allows input sampling rates of 32, 44.1 and 48 KHz and supports output bit rates of 32, 48, 56, 64, 96, 112, 128, 192, 256 and 384 Kbits/s. The transmission can be mono, dual channel (e.g. bilingual), stereo and joint stereo which is where advantage is taken of redundancy between the two audio channels.

MPEG Layer I is a simplified version of MUSICAM which is appropriate for the mild compression applications at low cost. It is very similar to PASC. Layer II is identical to MUSICAM and is very likely to be used for DAB. Layer III is a combination of the best features of ASPEC and MUSICAM and is mainly applicable to telecommunications where high compression factors are required.

The earlier MPEG-1 standard compresses audio and video into about 1.5 Mbits/s. The audio content of MPEG-1 may be used on its own to encode one or two channels at bit rates up to 448 Kbits/s. MPEG-2 allows the number of channels to increase to five: Left, Right, Centre, Left surround, Right surround and Subwoofer. In order to retain reverse compatibility with MPEG-1, the MPEG-2 coding converts the five-channel input to a compatible two-channel signal, L_o, R_o, by matrixing.[19] The data from these two channels is encoded in a standard MPEG-1 audio frame, and this is followed in MPEG-2 by an ancillary data frame which an MPEG-1 decoder will ignore. The ancillary frame contains data for any three of the audio channels. An MPEG-2 decoder will extract those three channels in addition to the MPEG-1 frame and then recover all five original channels by an inverse matrix.

In the USA, it has been proposed to use an alternative compression technique for the audio content of ATSC (Advanced Television Systems Committee) digital television broadcasts. This is the AC-3 system developed by Dolby Laboratories. The MPEG transport stream structure has also been standardized to allow it to carry AC-3 coded audio. The Digital Video Disc can also carry AC-3 or MPEG audio coding.

4.13   MPEG audio compression tools

There are many different approaches to audio compression, each having advantages and disadvantages. MPEG audio coding combines these tools in various ways in the three different coding levels. The approach of this chapter will be to examine the tools separately before seeing how they are used in MPEG and AC-3.

The simplest coding tool is companding which is a digital parallel of the analog noise reducers used in tape recording. Figure 4.15(a) shows that in companding the input signal level is monitored. Whenever the input level falls below maximum, it is amplified at the coder. The gain which was applied at the coder is added to the data stream so that the decoder can apply an equal attenuation. The advantage of companding is that the signal is kept as far away from the noise floor as possible. In analog noise reduction this is used to maximize the SNR of a tape recorder, whereas in digital compression it is used to keep the signal level as far as possible above the noises and artifacts introduced by various coding steps.

One common way of obtaining coding gain is to shorten the wordlength of samples so that fewer bits need to be transmitted. Figure 4.15(b) shows that when this is done, the noise floor will rise by 6 dB for every bit removed. This is because removing a bit halves the number of quantizing intervals which then must be twice as large, doubling the noise level. Clearly if this step follows the compander of (a), the audibility of the noise will be minimized. As an alternative to shortening the wordlength, the uniform quantized PCM signal can be converted to a non-uniform format. In non-uniform coding, shown in (c), the size of the quantizing step rises with the magnitude of the sample so that the noise level is greater when higher levels exist.

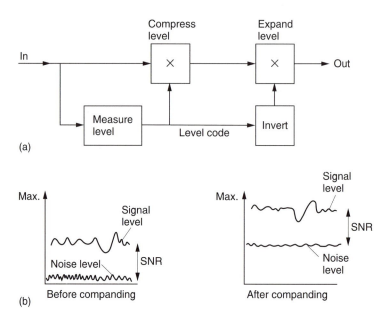

(a)

(b)

Figure 4.15 Digital companding. In (a) the encoder amplifies the input to maximum level and the decoder attenuates by the same amount. (b) In a companded system, the signal is kept as far as possible above the noise caused by shortening the sample wordlength.

Companding is a relative of floating-point coding shown in Figure 4.16 where the sample value is expressed as a mantissa and a binary exponent which determines how the mantissa needs to be shifted to have its correct absolute value on a PCM scale. The exponent is the equivalent of the gain setting or scale factor of a compandor.

Clearly in floating point the signal-to-noise ratio is defined by the number of bits in the mantissa, and, as shown in Figure 4.17, this will vary as a sawtooth function of signal level, as the best value, obtained when the mantissa is near overflow, is replaced by the worst value when the mantissa overflows and the exponent is incremented. Floating-point

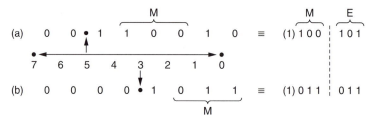

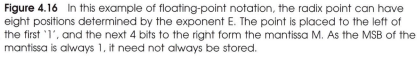

Figure 4.16 In this example of floating-point notation, the radix point can have eight positions determined by the exponent E. The point is placed to the left of the first '1', and the next 4 bits to the right form the mantissa M. As the MSB of the mantissa is always 1, it need not always be stored.

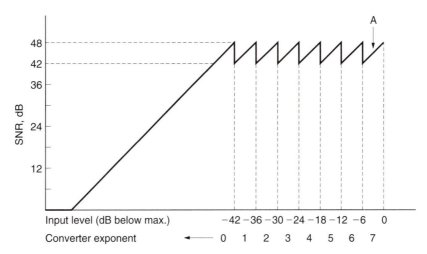

Figure 4.17 In this example of an 8-bit mantissa, 3-bit exponent system, the maximum SNR is 6 dB × 8 = 48 dB with maximum input of 0 dB. As input level falls by 6 dB. the converter noise remains the same, so SNR falls to 42 dB. Further reduction in signal level causes the converter to shift range (point A in the diagram) by increasing the input analog gain by 6 dB. The SNR is restored, and the exponent changes from 7 to 6 in order to cause the same gain change at the receiver. The noise modulation would be audible in this simple system. A longer mantissa word is needed in practice.

notation is used within DSP chips as it eases the computational problems involved in handling long wordlengths. For example, when multiplying floating-point numbers, only the mantissae need to be multiplied. The exponents are simply added.

A floating-point system requires one exponent to be carried with each mantissa and this is wasteful because in real audio material the level does not change so rapidly and there is redundancy in the exponents. A better alternative is floating-point block coding, also known as near-instantaneous companding, where the magnitude of the largest sample in a block is used to determine the value of an exponent which is valid for the whole block. Sending one exponent per block requires a lower data rate than in true floating point.[20]

In block coding the requantizing in the coder raises the quantizing noise, but it does so over the entire duration of the block. Figure 4.18 shows that if a transient occurs towards the end of a block, the decoder will reproduce the waveform correctly, but the quantizing noise will start at the beginning of the block and may result in a pre-noise (also called pre-echo) where the noise is audible before the transient. Temporal masking may be used to make this inaudible. With a 1 millisecond block, the artifacts are too brief to be heard.

Another solution is to use a variable time window according to the transient content of the audio waveform. When musical transients occur,

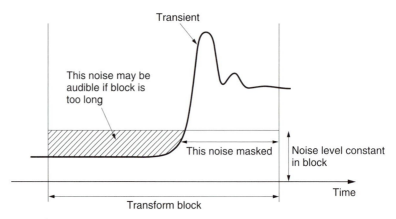

Figure 4.18   If a transient occurs towards the end of a transform block, the quantizing noise will still be present at the beginning of the block and may result in a pre-echo where the noise is audible before the transient.

short blocks are necessary and the coding gain will be low. At other times the blocks become longer allowing a greater coding gain.

While the above systems used alone do allow coding gain, the compression factor has to be limited because little benefit is obtained from masking. This is because the techniques above produce noise which spreads equally over the entire audio band. If the audio input spectrum is narrow, the noise will not be masked.

Sub-band coding splits the audio spectrum up into many different frequency bands. Once this has been done, each band can be individually processed. In real audio signals most bands will contain lower-level signals than the loudest one. Individual companding of each band will be more effective than broadband companding. Sub-band coding also allows the noise floor to be raised selectively so that noise is only added at frequencies where spectral masking will be effective.

There is little conceptual difference between a sub-band coder with a large number of bands and a transform coder. In transform coding, a Fourier or DCT transform of the waveform is computed periodically. Since the transform of an audio signal changes slowly, it needs be sent much less often than audio samples. The receiver performs an inverse transform. Transform coding is considered in section 4.14. Finally the data may be subject to a lossless binary compression using, for example, a Huffman code.

4.14   Transform coding

Audio is usually considered to be a time-domain waveform as this is what emerges from a microphone. As has been seen in Chapter 3, spectral analysis allows any periodic waveform to be represented by a set of harmonically related components of suitable amplitude and phase. In theory it is perfectly possible to decompose a periodic input waveform into its constituent frequencies and phases, and to record or transmit the transform. The transform can then be inverted and the original waveform will be precisely re-created.

Although one can think of exceptions, the transform of a typical audio waveform changes relatively slowly. The slow speech of an organ pipe or a violin string, or the slow decay of most musical sounds allow the rate at which the transform is sampled to be reduced, and a coding gain results. At some frequencies the level will be below maximum and a shorter wordlength can be used to describe the coefficient. Further coding gain will be achieved if the coefficients describing frequencies which will experience masking are quantized more coarsely. The transform of an audio signal is computed in the main signal path in a transform coder, and has sufficient frequency resolution to drive the masking model directly.

In practice there are some difficulties: real sounds are not periodic, but contain transients which transformation cannot accurately locate in time. The solution to this difficulty is to cut the waveform into short segments and then to transform each individually. The delay is reduced, as is the computational task, but there is a possibility of artifacts arising because of the truncation of the waveform into rectangular time windows. A solution is to use window functions (see Chapter 2) and to overlap the segments as shown in Figure 4.19. Thus every input sample appears in just two transforms, but with variable weighting depending upon its position along the time axis.

Figure 4.19 Transform coding can only be practically performed on short blocks. These are overlapped using window functions in order to handle continuous waveforms.

The DFT (discrete frequency transform) does not produce a continuous spectrum, but instead produces coefficients at discrete frequencies. The frequency resolution (i.e. the number of different frequency coefficients) is equal to the number of samples in the window. If overlapped windows are used, twice as many coefficients are produced as are theoretically necessary. In addition the DFT requires intensive computation, owing to the requirement to use complex arithmetic to render the phase of the components as well as the amplitude. An alternative is to use discrete cosine transforms (DCT).

4.15   MPEG Layer I audio coding

Figure 4.20 shows a block diagram of a Layer I coder which is a simplified version of that used in the MUSICAM system. A polyphase quadrature mirror filter network divides the audio spectrum into 32 equal sub-bands. The output data rate of the filter bank is no higher than the input rate because each band has been heterodyned to a frequency range from DC upwards.

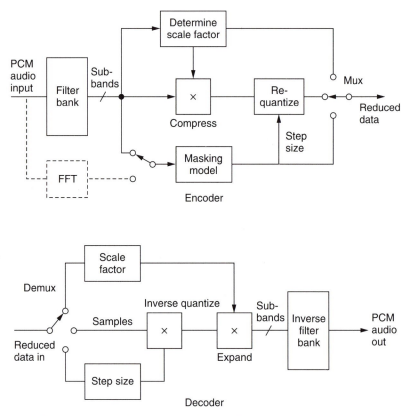

Figure 4.20 A simple sub-band coder. The bit allocation may come from analysis of the sub-band energy, or, for greater reduction, from a spectral analysis in a side chain.

Sub-band compression takes advantage of the fact that real sounds do not have uniform spectral energy. The wordlength of PCM audio is based on the dynamic range required and this is generally constant with frequency although any pre-emphasis will affect the situation. When a signal with an uneven spectrum is conveyed by PCM, the whole dynamic range is occupied only by the loudest spectral component, and all the other components are coded with excessive headroom. In its simplest form, sub-band coding[21] works by splitting the audio signal into a number of frequency bands and companding each band according to its own level. Bands in which there is little energy result in small amplitudes which can be transmitted with short wordlength. Thus each band results in variable-length samples, but the sum of all the sample wordlengths is less than that of PCM and so a coding gain can be obtained.

As MPEG audio coding relies on auditory masking, the sub-bands should preferably be narrower than the critical bands of the ear, hence the large number required. Figure 4.21 shows the critical condition where the masking tone is at the top edge of the sub-band. It will be seen that the narrower the sub-band, the higher the requantizing noise that can be masked. The use of an excessive number of sub-bands will, however, raise complexity and the coding delay.

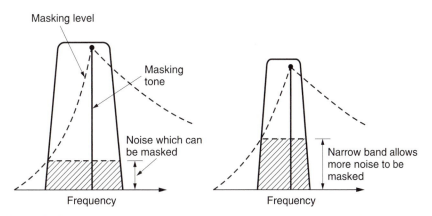

Figure 4.21   In sub-band coding the worst case occurs when the masking tone is at the top edge of the sub-band. The narrower the band, the higher the noise level which can be masked.

Constant-size input blocks are used, containing 384 samples. At 48 KHz, 384 samples corresponds to a period of 8 milliseconds. After the sub-band filter each band contains 12 samples per block. The block size was based on the pre-masking phenomenon of Figure 4.18. The samples in each sub-band block or *bin* are companded according to the peak value in the bin. A 6-bit scale factor is used for each sub-band which applies to all 12 samples.

If a fixed compression factor is employed, the size of the coded output block will be fixed. The wordlengths in each bin will have to be such that the sum of the bits from all the sub-band equals the size of the coded block. Thus some sub-bands can have long wordlength coding if others have short wordlength coding. The process of determining the requantization step size, and hence the wordlength, in each sub-band is known as bit allocation.

For simplicity, in Layer I the levels in the 32 sub-bands themselves are used as a crude spectral analysis of the input in order to drive the masking model. The masking model uses the input spectrum to determine a new threshold of hearing which in turn determines how much the noise floor can be raised in each sub-band. Where masking takes place, the signal is quantized more coarsely until the quantizing noise is raised to just below the masking level. The coarse quantization requires shorter wordlengths and allows a coding gain. The bit allocation may be iterative as adjustments are made to obtain the best NMR within the allowable data rate.

The samples of differing wordlength in each bin are then assembled into the output coded block. Unlike a PCM block, which contains samples of fixed wordlength, a coded block contains many different wordlengths and these can vary from one block to the next. In order to deserialize the block into samples of various wordlength and demultiplex the samples into the appropriate frequency bins, the decoder has to be told what bit allocations were used when it was packed, and some synchronizing means is needed to allow the beginning of the block to be identified.

The compression factor is determined by the bit-allocation system. It is not difficult to change the output block size parameter to obtain a different compression factor. If a larger block is specified, the bit allocator simply iterates until the new block size is filled. Similarly the decoder need only deserialize the larger block correctly into coded samples and then the expansion process is identical except for the fact that expanded words contain less noise. Thus codecs with varying degrees of compression are available which can perform different bandwidth/performance tasks with the same hardware.

Figure 4.22 shows the format of the Layer I data stream. The frame begins with a sync pattern to reset the phase of deserialization, and a header which describes the sampling rate and any use of pre-emphasis. Following this is a block of 32 4-bit allocation codes. These specify the wordlength used in each sub-band and allow the decoder to deserialize the sub-band sample block. This is followed by a block of 32 6-bit scale factor indices, which specify the gain given to each band during companding. The last block contains 32 sets of 12 samples. These samples vary in wordlength from one block to the next, and can be from 0 to 15 bits long. The deserializer has to use the 32 allocation information codes to work out how to deserialize the sample block into individual samples of variable length.

The Layer I MPEG decoder is shown in Figure 4.23. The elementary stream is deserialized using the sync pattern and the variable-length samples are asembled using the allocation codes. The variable length

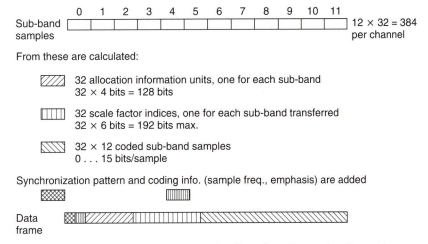

From these are calculated:

32 allocation information units, one for each sub-band
32 × 4 bits = 128 bits

32 scale factor indices, one for each sub-band transferred
32 × 6 bits = 192 bits max.

32 × 12 coded sub-band samples
0 . . . 15 bits/sample

Synchronization pattern and coding info. (sample freq., emphasis) are added

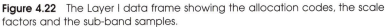

Figure 4.22 The Layer I data frame showing the allocation codes, the scale factors and the sub-band samples.

samples are returned to 15-bit wordlength by adding zeros. The scale factor indices are then used to determine multiplication factors used to return the waveform in each sub-band to its original level. The 32 sub-band signals are then merged into one spectrum by the synthesis filter. This is a set of bandpass filters which returns every sub-band to the correct place in the audio spectrum and then adds them to produce the audio output.

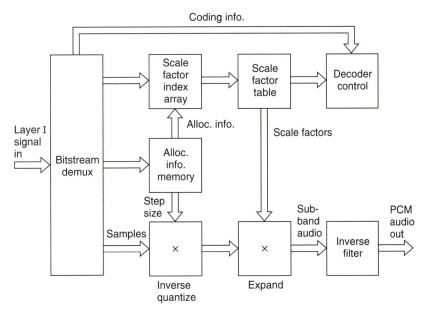

Figure 4.23 The Layer I decoder. See text for details.

4.16   MPEG Layer II audio coding

MPEG Layer II audio coding is identical to MUSICAM. The same 32-band filterbank and the same block companding scheme as Layer I is used. Figure 4.24 shows that using the level in a sub-band to drive the masking model is sub-optimal because it is not known where in the sub-band the energy lies. As the skirts of the masking curve are asymmetrical, the noise floor can be raised higher if the masker is at the low end of the sub-band than if it is at the high end.

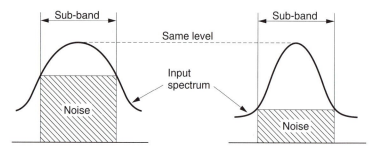

Figure 4.24   Accurate knowledge of the spectrum allows the noise floor to be raised higher while remaining masked.

In order to give better spectral resolution than the filterbank, a side chain FFT is computed having 1024 points, resulting in an analysis of the audio spectrum eight times better than the sub-band width. The FFT drives the masking model which controls the bit allocation. In order to give the FFT sufficient resolution, the block length is increased to 1152 samples. This is three times the block length of Layer I.

The QMF bandsplitting technique is restricted to bands of equal width. It might be thought that this is a drawback because the critical bands of the ear are non-uniform. In fact this is only a problem when very low bit rates are required. In all cases it is the masking model of hearing which must have correct critical bands. This model can then be superimposed on bands of any width to determine how much masking and therefore coding gain is possible. Uniform-width sub-bands will not be able to obtain as much masking as bands which are matched to critical bands, but for many applications the additional coding gain is not worth the added filter complexity.

The block companding scheme of Layer II is the same as in Layer I because the 1152-sample block is divided into three 384-sample blocks. However, not all the scale factors are transmitted, because they contain a degree of redundancy on real program material. The difference between scale factors in successive blocks in the same band exceeds 2 dB less than

10% of the time. Layer II analyses the set of three successive scale factors in each sub-band. On stationary program, these will be the same and only one scale factor out of three is sent. As the transient content increases in a given sub-band, two or three scale factors will be sent. A scale factor select code must be sent to allow the decoder to determine what has been sent in each sub-band. This technique effectively halves the scale factor bit rate.

The requantized samples in each sub-band, bit-allocation data, scale factors and scale factor select codes are multiplexed into the output bit stream.

The Layer II decoder is not much more complex than the Layer I decoder as the only additional processing is to decode the compressed scale factors to produce one scale factor per 384-sample block.

4.17   MPEG Layer III audio coding

This is the most complex layer of the ISO standard, and is only really necessary when the most severe data rate constraints must be met with high quality. It is a transform code based on the ASPEC system with certain modifications to give a degree of commonality with Layer II. The original ASPEC coder used a direct MDCT on the input samples. In Layer III this was modified to use a hybrid transform incorporating the existing polyphase 32-band QMF of Layers I and II. In Layer III, the 32 sub-bands from the QMF are each processed by a 12-band MDCT to obtain 384 output coefficients. Two window sizes are used to avoid pre-echo on transients. The window switching is performed by the psychoacoustic model. It has been found that pre-echo is associated with the entropy in the audio rising above the average value.

A highly accurate perceptive model is used to take advantage of the high-frequency resolution available. Non-uniform quantizing is used, along with Huffman coding. This is a technique where the most common code values are allocated the shortest wordlength.

4.18   Dolby AC-3

Dolby AC-3 is in fact a family of transform coders based on time-domain aliasing cancellation (TDAC) which allow various compromises between coding delay and bit rate to be used. In the modified discrete cosine transform (MDCT),[22] windows with 50% overlap are used. Thus twice as many coefficients as necessary are produced. These are sub-sampled by a factor of two to give a critically sampled transform, which results in potential aliasing in the frequency domain. However, by making a slight change to the transform, the alias products in the second half of a given window are equal in size but of opposite polarity to the alias products in the first half of the next window, and so will be cancelled on reconstruction. This is the principle of TDAC.

Figure 4.25 shows the generic block diagram of the AC-3 coder. Input audio is divided into 50% overlapped blocks of 512 samples. These are subject to a TDAC transform which uses alternate modified sine and cosine transforms. The transforms produce 512 coefficients per block, but these are redundant and after the redundancy has been removed there are 256 coefficients per block.

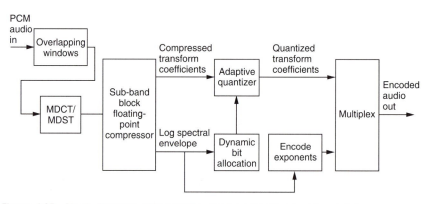

Figure 4.25 Block diagram of the Dolby AC-3 coder, See text for details.

The input waveform is constantly analysed for the presence of transients and if these are present the block length will be halved to prevent pre-noise. This halves the frequency resolution but doubles the temporal resolution.

The coefficients have high-frequency resolution and are selectively combined in sub-bands which approximate the critical bands. Coefficients in each sub-band are normalized and expressed in floating-point block notation with common exponent. The exponents in fact represent the logarithmic spectral envelope of the signal and can be used to drive the perceptive model which operates the bit allocation. The mantissae of the transform coefficients are then requantized according to the bit allocation.

The output bit stream consists of the requantized coefficients and the log spectral envelope in the shape of the exponents. There is a great deal of redundancy in the exponents. In any block, only the first exponent, corresponding to the lowest frequency, is transmitted absolutely. Remaining coefficients are transmitted differentially. Where the input has a smooth spectrum the exponents in several bands will be the same and the differences will then be zero. In this case exponents can be grouped using flags.

Further use is made of temporal redundancy. An AC-3 sync frame contains six blocks. The first block of the frame contains absolute exponent data, but where stationary audio is encountered, successive blocks in the frame can use the same exponents.

The receiver uses the log spectral envelope to deserialize the mantissae of the coefficients into the correct wordlengths. The highly redundant exponents are decoded starting with the lowest frequency coefficient in the first block of the frame and adding differences to create the remainder. The exponents are then used to return the coefficients to fixed-point notation. Inverse transforms are then computed, followed by a weighted overlapping of the windows to obtain PCM data.

4.19   Compression in stereo

Once hardware became economic, the move to digital audio was extremely rapid. One of these reasons was the sheer sound quality available. When well engineered, the PCM digital domain does so little damage to the sound quality that the problems of the remaining analog parts usually dominate. The one serious exception to this is lossy compression which does not preserve the original waveform and must therefore be carefully assessed before being used in high-quality applications.

In a monophonic system, all the sound is emitted from a single point and psychoacoustic masking operates to its fullest extent. Audio compression techniques of the kind described above work well in mono. However, in stereophonic (which in this context also includes surround sound) applications different criteria apply. In addition to the timbral information describing the nature of the sound sources, stereophonic systems also contain spatial information describing their location.

The greatest cause for concern is that in stereophonic systems, masking is not as effective. When two sound sources are in physically different locations, the degree of masking is not as great as when they are co-sited. Unfortunately all the psychoacoustic masking models used in today's compressors assume co-siting. When used in stereo or surround systems, the artifacts of the compression process can be revealed. This was first pointed out by the late Michael Gerzon who introduced the term *unmasking* to describe the phenomenon.

The hearing mechanism has an ability to concentrate on one of many simultaneous sound sources based on direction. The brain appears to be able to insert a controllable time delay in the nerve signals from one ear with respect to the other so that when sound arrives from a given direction the nerve signals from both ears are coherent causing the binaural threshold of hearing to be 3–6 dB better than monaural at around 4 KHz. Sounds arriving from other directions are incoherent and are heard less well. This is known as *attentional selectivity*, or more colloquially as the *cocktail party effect*.[23]

Human hearing can locate a number of different sound sources simultaneously by constantly comparing excitation patterns from the two ears with different delays. Strong correlation will be found where the delay corresponds to the interaural delay for a given source. This delay-varying mechanism will take time and the ear is slow to react to changes in source direction. Oscillating sources can only be tracked up to 2–3 Hz and the ability to locate bursts of noise improves with burst duration up to about 700 milliseconds.

Monophonic systems prevent the use of either of these effects completely because the first version of all sounds reaching the listener comes from the same loudspeaker. Stereophonic systems allow attentional selectivity to function in that the listener can concentrate on specific sound sources in a reproduced stereophonic image with the same facility as in the original sound.

When two sound sources are spatially separated, if the listener uses attentional selectivity to concentrate on one of them, the contributions from both ears will correlate. This means that the contributions from the other sound will be decorrelated, reducing its masking ability significantly.

Experiments showed long ago that even technically poor stereo was always preferred to pristine mono. This is because we are accustomed to sounds and reverberation coming from all different directions in real life and having them all superimposed in a mono speaker convinces no-one, however accurate the waveform.

We live in a reverberant world which is filled with sound reflections. If we could separately distinguish every different reflection in a reverberant room we would hear a confusing cacophony. In practice we hear very well in reverberant surroundings, far better than microphones can, because of the transform nature of the ear and the way in which the brain processes nerve signals. Because the ear has finite frequency discrimination ability in the form of critical bands, it must also have finite temporal discrimination.

When two or more versions of a sound arrive at the ear, provided they fall within a time span of about 30 milliseconds, they will not be treated as separate sounds, but will be fused into one sound. Only when the time separation reaches 50–60 milliseconds do the delayed sounds appear as echoes from different directions. As we have evolved to function in reverberant surroundings, most reflections do not impair our ability to locate the source of a sound.

Clearly the first version of a transient sound to reach the ears must be the one which has travelled by the shortest path and this must be the direct sound rather than a reflection. Consequently the ear has evolved to attribute source direction from the time of arrival difference at the two ears of the first version of a transient. This phenomenon is known as the precedence effect.

Intensity stereo, the type of signal format obtained with coincident microphones or panpots, works purely by amplitude differences at the two loudspeakers. The two signals should be exactly in phase. As both ears hear both speakers the result is that the space between the speakers and the ears turns the intensity differences into time of arrival differences. These give the illusion of virtual sound sources.

A virtual sound source from a panpot has zero width and on ideal loudspeakers would appear as a virtual point source. Figure 4.26(a) shows how a panpotted dry mix should appear spatially on ideal speakers whereas (b) shows what happens when artificial stereo reverb is added. Figure 4.26(b) is also what is obtained with real sources using a coincident pair of high-quality mikes. In this case the sources are the real sources and the sound between is reverb/ambience.

When listening on high-quality loudspeakers, audio compressors change the characteristic of Figure 4.26(b) to that shown in (c). Even at high bit rates, corresponding to the smallest amount of compression, there is an audible difference between the original and the compressed result. The dominant sound sources are reproduced fairly accurately, but what is most striking is that the ambience and reverb between is dramatically reduced or absent, making the decoded sound much drier than the original. It will also be found that the rate of decay of reverberation is accelerated as shown in (d).

These effects are heard because the reverberation exists at a relatively low level. The coder will assume that it is inaudible due to masking and remove or attenuate it. The effect is apparent to the same extent with both MPEG Layer II and Dolby AC-3 coders even though their internal workings are quite different. This is not surprising because both are probably based on the same psychoacoustic masking model.

MPEG Layer III fares quite badly in stereo because the bit rate is lower. Transient material has a peculiar effect whereby the ambience would come and go according to the entropy of the dominant source. A percussive note would narrow the sound stage and appear dry but afterwards the reverb level would come back up. The effects are not subtle and do not require 'golden ears' or special source material. The author has successfully demonstrated these effects under conference conditions. All these effects disappear when the signals to the speakers are added to make mono as this prevents attentional selectivity and unmasking cannot occur.

The above compression artifacts are less obvious or even absent when poor-quality loudspeakers are used. Loudspeakers are part of the

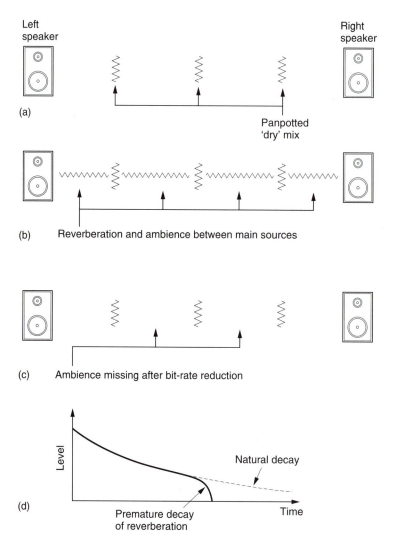

Figure 4.26 Compression is less effective in stereo. In (a) is shown the spatial result of a 'dry' panpotted mix. (b) shows the result after artificial reverberation which can also be obtained in an acoustic recording with coincident mikes. After compression (c) the ambience and reverberation may be reduced or absent. (d) Reverberation may also decay prematurely.

communication channel and have an information capacity, both timbral and spatial. If the quality of the loudspeaker is poor, it may remove more information from the signal than the preceding compressor and coding artifacts will not be revealed.

The precedence effect allows the listener to locate the source of a sound by concentrating on the first version of the sound and rejecting later versions. Versions which may arrive from elsewhere simply add to the perceived loudness but do not change the perceived location of the source.

The precedence effect can only reject reverberant sounds which arrive after the inter-aural delay. When reflections arrive within the inter-aural delay of about 700 microseconds the precedence effect breaks down and the perceived direction can be pulled away from that of the first arriving source by an increase in level. Figure 4.27 shows that this area is known as the time–intensity trading region. Once the maximum inter-aural delay is exceeded, the hearing mechanism knows that the time difference must be due to reverberation and the trading ceases to change with level.

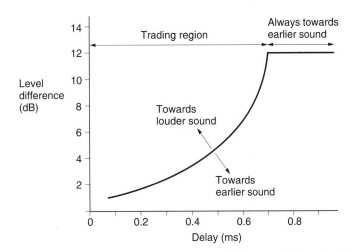

Figure 4.27 Time–intensity trading occurs within the inter-aural delay period.

Unfortunately reflections with delays of the order of 700 microseconds are exactly what are provided by the traditional rectangular loudspeaker with flat sides and sharp corners. The discontinuities between the panels cause impedance changes which act as acoustic reflectors. The loudspeaker becomes a multiple source producing an array of signals within the time–intensity trading region and instead of acting as a point source the loudspeaker becomes a distributed source.

Figure 4.28 shows that when a loudspeaker is a distributed source, it cannot create a point image. Instead the image is also distributed, an effect known as smear. Note that the point sources have spread so that there are almost no gaps between them, effectively masking the ambience. If a compressor removes it the effect cannot be heard. It may erroneously be assumed that the compressor is transparent when in fact it is not.

For the above reasons, the conclusions of many of the listening tests carried out on audio compressors are simply invalid. The author has read descriptions of many such tests, and in almost all cases little is said about the loudspeakers used or about what steps were taken to measure their

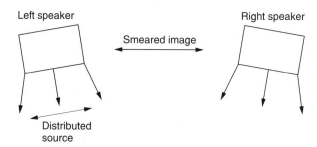

Figure 4.28 A loudspeaker which is a distributed source cannot produce a point stereo image; only a spatially spread or smeared image.

spatial accuracy. This is simply unscientific and has led to the use of compression being more widespread than is appropriate.

A type of flat panel loudspeaker which uses chaotic bending modes of the diaphragm has recently been introduced. It should be noted that these speakers are distributed sources and and so are quite unsuitable for assessing compression quality.

While compression may be adequate to deliver post-produced audio to a consumer with mediocre loudspeakers, these results underline that it has no place in a high-quality production environment. When assessing codecs, loudspeakers having poor diffraction design will conceal artefacts. When mixing for a compressed delivery system, it will be necessary to include the codec in the monitor feeds so that the results can be compensated. Where high-quality stereo is required, either full bit rate PCM or lossless (packing) techniques must be used.

An interesting corollary of the above is that compressors can be used to test stereo loudspeakers. The arrangement of Figure 4.29 is used. Starting with the highest bit rate possible, the speakers are switched between the source material and the output of the codec. The compression factor is increased until the effects noted above are heard. When the effects are just audible, the bit rate of the compressor is at the information capacity of the loudspeakers.

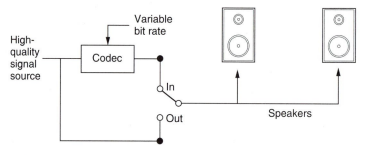

Figure 4.29 Using compression to test loudspeakers. The better the loudspeaker, the less compression can be used before it becomes audible.

Domestic bookshelf loudspeakers measure approximately 100 Kbits/s. Traditional square box hi-fi speakers measure around 200 Kbits/s, but a good pair of electrostatics will go above 500 Kbits/s.

References

1. Johnston, J.D., Transform coding of audio signals using perceptual noise criteria. *IEEE J. Selected Areas in Comms.*, **JSAC-6**, 314–323 (1988)
2. Moore, B.C.J., *An Introduction to the Psychology of Hearing*, London: Academic Press (1989)
3. Muraoka, T., Iwahara, M. and Yamada, Y., Examination of audio bandwidth requirements for optimum sound signal transmission. *J. Audio Eng. Soc.*, **29**, 2–9 (1982)
4. Muraoka, T., Yamada, Y. and Yamazaki, M., Sampling frequency considerations in digital audio. *J. Audio Eng. Soc.*, **26**, 252–256 (1978)
5. Fincham, L.R., The subjective importance of uniform group delay at low frequencies. Presented at the 74th Audio Engineering Society Convention (New York, 1983), Preprint 2056(H-1)
6. Fletcher, H., Auditory patterns. *Rev. Modern Physics*, **12**, 47–65 (1940)
7. Zwicker, E., Subdivision of the audible frequency range into critical bands. *J. Acoust. Soc. Amer.*, **33**, 248 (1961)
8. Moore, B. and Glasberg, B., Formulae describing frequency selectivity as a function of frequency and level, and their use in calculating excitation patterns. *Hearing Research*, **28**, 209–225 (1987)
9. Grewin, C. and Ryden, T., Subjective assessments on low bit-rate audio codecs. *Proc. 10th. Int. Audio Eng. Soc. Conf.*, 91–102, New York: Audio Eng. Soc. (1991)
10. Gilchrist, N.H.C., Digital sound: the selection of critical programme material and preparation of the recordings for CCIR tests on low bit rate codecs. *BBC Res. Dept. Rep.*, RD 1993/1
11. Colomes, C. and Faucon, G., A perceptual objective measurement system (POM) for the quality assessment of perceptual codecs. Presented at 96th Audio Eng. Soc. Conv. Amsterdam (1994), Preprint No. 3801 (P4.2)
12. Johnston, J., Estimation of perceptual entropy using noise masking criteria. *ICASSP*, 2524–2527 (1988)
13. Gilchrist, N.H.C., Delay in broadcasting operations. Presented at 90th Audio Eng. Soc. Conv. (1991), Preprint 3033
14. Wiese, D., MUSICAM: flexible bitrate reduction standard for high quality audio. Presented at Digital Audio Broadcasting Conference (London, March 1992)
15. Brandenburg, K., ASPEC coding. *Proc. 10th. Audio Eng. Soc. Int. Conf.*, 81–90, New York: Audio Eng. Soc. (1991)
16. ISO/IEC JTC1/SC2/WG11 N0030: MPEG/AUDIO test report, Stockholm (1990)
17. ISO/IEC JTC1/SC2/WG11 MPEG 91/010, The SR report on The MPEG/AUDIO subjective listening test. Stockholm (1991)
18. ISO/IEC JTC1/SC2/WG11, Committee draft 11172
19. Bonicel, P. *et al.*, A real time ISO/MPEG2 multichannel decoder. Presented at 96th Audio Eng. Soc. Conv. (1994), Preprint No. 3798 (P3.7)4.30
20. Caine, C.R., English, A.R. and O'Clarey, J.W.H., NICAM-3: near-instantaneous companded digital transmission for high-quality sound programmes. *J. IERE*, **50**, 519–530 (1980)
21. Crochiere, R.E., Sub-band coding. *Bell System Tech. J.*, **60**, 1633–1653 (1981)
22. Princen, J.P., Johnson, A. and Bradley, A.B., Sub-band/transform coding using filter bank designs based on time domain aliasing cancellation. *Proc. ICASSP*, 2161–2164 (1987)
23. Watkinson, J.R., *The Art of Sound Reproduction*, Chapter 7, Oxford: Focal Press (1988)

5

MPEG-2 video compression

In this chapter the principles of video compression are explored, leading to descriptions of MPEG-2. As MPEG-2 supports both progressive and interlaced scan, this chapter uses the term 'picture' to mean an image of any kind at one point on the time axis. This could be a field or a frame in interlaced systems but only a frame in non-interlaced systems. The terms field and frame will be used only when the distinction is important.

5.1   The eye

All television signals ultimately excite some response in the eye and the viewer can only describe the result subjectively. Familiarity with the functioning and limitations of the eye is essential to an understanding of video compression.

The simple representation of Figure 5.1 shows that the eyeball is nearly spherical and is swivelled by muscles. The space between the cornea and the lens is filled with transparent fluid known as *aqueous humour*. The remainder of the eyeball is filled with a transparent jelly known as *vitreous humour*.

Light enters the cornea, and the amount of light admitted is controlled by the pupil in the iris. Light entering is involuntarily focused on the retina by the lens in a process called *visual accommodation*. The lens is the only part of the eye which is not nourished by the bloodstream and its centre is technically dead. In a young person the lens is flexible and muscles distort it to perform the focusing action. In old age the lens loses some flexibility and causes *presbyopia* or limited accommodation. In some people the length of the eyeball is incorrect resulting in *myopia* (short-sightedness) or *hypermetropia*

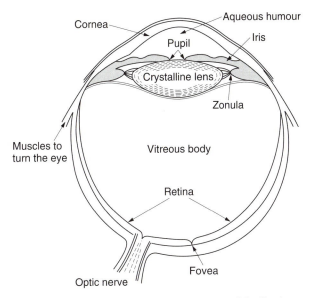

Figure 5.1 The eyeball is in effect a living camera except for the lens which receives no blood supply and is technically dead.

(long-sightedness). The cornea should have the same curvature in all meridia, and if this is not the case, *astigmatism* results.

The retina is responsible for light sensing and contains a number of layers. The surface of the retina is covered with arteries, veins and nerve fibres and light has to penetrate these in order to reach the sensitive layer. This contains two types of discrete receptors known as *rods* and *cones* from their shape. The distribution and characteristics of these two receptors are quite different. Rods dominate the periphery of the retina whereas cones dominate a central area known as the *fovea* outside which their density drops off. Vision using the rods is monochromatic and has poor resolution but remains effective at very low light levels, whereas the cones provide high resolution and colour vision but require more light. Figure 5.2 shows how the sensitivity of the retina slowly increases in response to entering darkness. The first part of the curve is the adaptation of cone or *photopic* vision. This is followed by the greater adaptation of the rods in *scotopic* vision. At such low light levels the fovea is essentially blind and small objects which can be seen in the peripheral rod vision disappear when stared at.

The cones in the fovea are densely packed and directly connected to the nervous system allowing the highest resolution. Resolution then falls off away from the fovea. As a result the eye must move to scan large areas of detail. The image perceived is not just a function of the retinal response, but is also affected by processing of the nerve signals. The overall acuity of the eye can be displayed as a graph of the response plotted against the

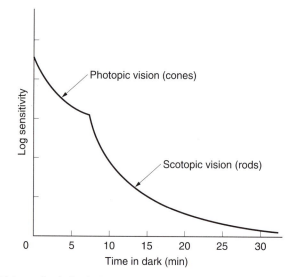

Figure 5.2 Vision adapts to darkness in two stages known as photopic and scotopic vision.

degree of detail being viewed. Detail is generally measured in lines per millimetre or cycles per picture height, but this takes no account of the distance from the eye. A better unit for eye resolution is one based upon the subtended angle of detail as this will be independent of distance. Units of cycles per degree are then appropriate. Figure 5.3 shows the response of the eye to static detail. Note that the response to very low frequencies is also attenuated. An extension of this characteristic allows

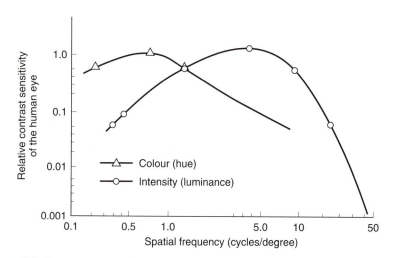

Figure 5.3 The response of the eye to static detail falls off at both low and high spatial frequencies.

the vision system to ignore the fixed pattern of shadow on the retina due to the nerves and arteries.

The retina does not respond instantly to light, but requires between 0.15 and 0.3 s before the brain perceives an image. The resolution of the eye is primarily a spatio-temporal compromise. The eye is a spatial sampling device; the spacing of the rods and cones on the retina represents a spatial sampling frequency.

The measured acuity of the eye exceeds the value calculated from the sample site spacing because a form of oversampling is used. The eye is in a continuous state of unconscious vibration called saccadic motion. This causes the sampling sites to exist in more than one location, effectively increasing the spatial sampling rate provided there is a temporal filter which is able to integrate the information from the various different positions of the retina.

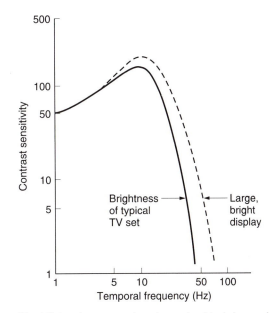

Figure 5.4 The critical flicker frequency is not constant but rises as brightness increases.

This temporal filtering is responsible for 'persistence of vision'. Flashing lights are perceived to flicker only until the critical flicker frequency (CFF) is reached; the light appears continuous for higher frequencies. The CFF is not constant but changes with brightness (see Figure 5.4). Note that the field rate of European television at 50 fields per second is marginal with bright images. Figure 5.5 shows the two-dimensional or spatio-temporal response of the eye.

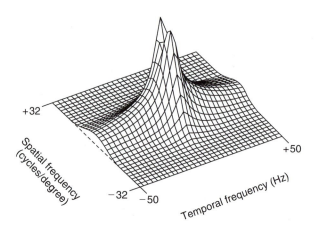

Figure 5.5 The response of the eye shown with respect to temporal and spatial frequencies. Note that even slow relative movement causes a serious loss of resolution. The eye tracks moving objects to prevent this loss.

If the eye were static, a detailed object moving past it would give rise to temporal frequencies, as Figure 5.6(a) shows. The temporal frequency is given by the detail in the object, in lines per millimetre, multiplied by the speed. Clearly a highly detailed object can reach high temporal frequencies even at slow speeds, yet Figure 5.5 shows that the eye cannot respond to high temporal frequencies.

However, the human viewer has an interactive visual system which causes the eyes to track the movement of any object of interest. Figure 5.6(b) shows that when eye tracking is considered, a moving object is rendered stationary with respect to the retina so that temporal frequencies fall to zero and much the same acuity to detail is available despite motion. This is known as dynamic resolution and it's how humans judge the detail in real moving pictures. It is somewhat surprising that video engineers so often state that softening of moving objects is inevitable and acceptable, when it plainly isn't.

5.2   Dynamic resolution

As the eye uses involuntary tracking at all times, the criterion for measuring the definition of moving image portrayal systems has to be *dynamic resolution*, defined as the apparent resolution perceived by the viewer in an object moving within the limits of accurate eye tracking. The traditional metric of static resolution in film and television has to be abandoned as unrepresentative.

Figure 5.7(a) shows that when the moving eye tracks an object on the screen, the viewer is watching with respect to the optic flow axis, not the

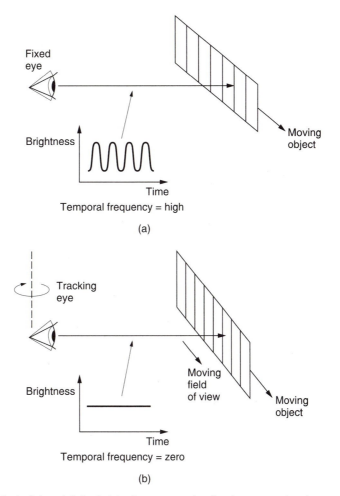

Figure 5.6 In (a) a detailed object moves past a fixed eye, causing temporal frequencies beyond the response of the eye. This is the cause of motion blur. In (b) the eye tracks the motion and the temporal frequency becomes zero. Motion blur cannot then occur.

time axis, and these are not parallel when there is motion. The optic flow axis is defined as an imaginary axis in the spatio-temporal volume which joins the same points on objects in successive frames. Clearly when many objects move independently there will be one optic flow axis for each.

The optic flow axis is identified by motion-compensated standards convertors to eliminate judder and also by MPEG compressors because the greatest similarity from one picture to the next is along that axis. The success of these devices is testimony to the importance of the theory.

According to sampling theory, a sampling system cannot properly convey frequencies beyond half the sampling rate. If the sampling rate is considered to be the picture rate, then no temporal frequency of more

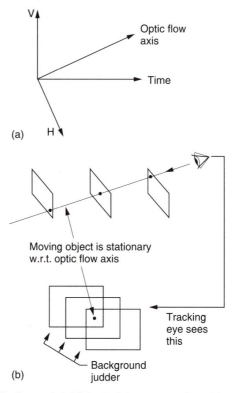

Figure 5.7 The optic flow axis (a) joins points on a moving object in successive pictures. (b) When a tracking eye follows a moving object on a screen, that screen will be seen in a different place at each picture. This is the origin of background strobing.

than 25 or 30 Hz can be handled (12 Hz for film). With a stationary camera and scene, temporal frequencies can only result from the brightness of lighting changing, but this will not approach the limit. However, when there is relative movement between camera and scene, detailed areas develop high temporal frequencies, just as was shown in Figure 5.6(a) for the eye. This is because relative motion results in a given point on the camera sensor effectively scanning across the scene. The temporal frequencies generated are beyond the limit set by sampling theory, and aliasing should take place. However, when the resultant pictures are viewed by a human eye, this aliasing is not perceived because, once more, the eye tracks the motion of the scene.[1]

Figure 5.8 shows what happens when the eye follows correctly. Although the camera sensor and display are both moving through the field of view, the original scene and the retina are now stationary with respect to one another. As a result the temporal frequency at the eye due to the object being followed is brought to zero and no aliasing is perceived by the viewer due to sampling in the image portrayal system. While this result is highly desirable, it does not circumvent sampling theory because the effect only works if several assumptions are made, including the requirement for the motion to be smooth.

Figure 5.7(b) shows that when the eye is tracking, successive pictures appear in different places with respect to the retina. In other words if an object is moving down the screen and followed by the eye, the raster is actually moving up with respect to the retina. Although the tracked object is stationary with respect to the retina and temporal frequencies are zero, the object is moving with respect to the sensor and the display and in those units high temporal frequencies will exist. If the motion of the object on the sensor is not correctly portrayed, dynamic resolution will suffer.

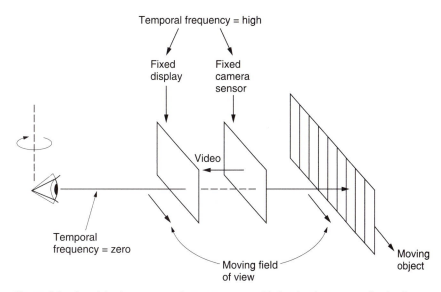

Figure 5.8 An object moves past a camera, and is tracked on a monitor by the eye. The high temporal frequencies cause aliasing in the TV signal, but these are not perceived by the tracking eye as this reduces the temporal frequency to zero.

In real-life eye tracking, the motion of the background will be smooth, but in an image-portrayal system based on periodic presentation of frames, the background will be presented to the retina in a different position in each frame. The retina seperately perceives each impression of the background leading to an effect called *background strobing*.

The criterion for the selection of a display frame rate in an imaging system is sufficient reduction of background strobing. It is a complete myth that the display rate simply needs to exceed the critical flicker frequency. Manufacturers of graphics displays which use frame rates well in excess of those used in film and television are doing so for a valid reason: it gives better results! Note that the display rate and the transmission rate need not be the same in an advanced system.

Dynamic resolution analysis confirms that both interlaced television and conventionally projected cinema film are both seriously sub-optimal. In contrast, progressively scanned television systems have no such defects.

5.3   Contrast

The contrast sensitivity of the eye is defined as the smallest brightness difference which is visible. In fact the contrast sensitivity is not constant, but increases proportionally to brightness. Thus whatever the brightness of an object, if that brightness changes by about 1% it will be equally detectable.

The true brightness of a television picture can be affected by electrical noise on the video signal. As contrast sensitivity is proportional to brightness, noise is more visible in dark picture areas than in bright areas. In practice the gamma characteristic of the CRT is put to good use in making video noise less visible. Instead of having linear video signals which are subjected to an inverse gamma function immediately prior to driving the CRT, the inverse gamma correction is performed at the camera. In this way the video signal is non-linear for most of its journey.

Figure 5.9 shows a reverse gamma function. As a true power function requires infinite gain near black, a linear segment is substituted. It will be

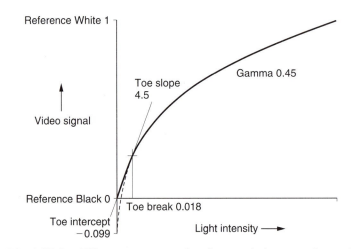

Figure 5.9 CCIR Rec.709 reverse gamma function used at camera has a straight line approximation at the lower part of the curve to avoid boosting camera noise. Note that the output amplitude is greater for modulation near black.

seen that contrast variations near black result in larger signal amplitude than variations near white. The result is that noise picked up by the video signal has less effect on dark areas than bright areas. After the gamma of the CRT has acted, noise near black is compressed with respect to noise near white. Thus a video transmission system using gamma correction at source has a better perceived noise level than if the gamma correction is performed near the display.

In practice the system is not rendered perfectly linear by gamma correction and a slight overall exponential effect is usually retained in order further to reduce the effect of noise in darker parts of the picture. A gamma-correction factor of 0.45 may be used to achieve this effect. If another type of display is to be used with signals designed for CRTs, the gamma characteristic of that display will probably be different and some gamma conversion will be required.

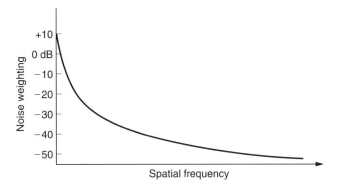

Figure 5.10 The sensitivity of the eye to noise is greatest at low frequencies and drops rapidly with increasing frequency. This can be used to mask quantizing noise caused by the compression process.

Sensitivity to noise is also a function of spatial frequency. Figure 5.10 shows that the sensitivity of the eye to noise falls with frequency from a maximum at zero. Thus it is vital that the average brightness should be correctly conveyed by a compression system, whereas higher spatial frequencies can be subject to more requantizing noise. Transforming the image into the frequency domain allows this characteristic to be explored.

5.4   Colour vision

Colour vision is due to the cones on the retina which occur in three different types, responding to different colours. Figure 5.11 shows that human vision is restricted to range of light wavelengths from 400

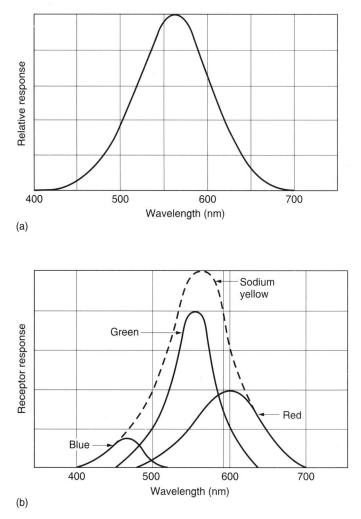

Figure 5.11 (a) The response of the eye to various wavelengths. There is a pronounced peak in the green region. (b) The different types of cone in the eye have the approximate responses shown here which allow colour vision.

nanometres to 700 nanometres. Shorter wavelengths are called ultra-violet and longer wavelengths are called infra-red. Note that the response is not uniform, but peaks in the area of green. The response to blue is very poor and makes a nonsense of the use of blue lights on emergency vehicles which owe much to tradition and little to psycho-optics.

Figure 5.11 shows an approximate response for each of the three types of cone. If light of a single wavelength is observed, the relative responses of the three sensors allows us to discern what we call the colour of the light. Note that at both ends of the visible spectrum there are areas in which only one receptor responds; all colours in those areas look the same. There is a great deal of variation in receptor response from one individual to the next and the curves used in television are the average of a great many tests. In a surprising number of people the single receptor zones are extended and discrimination between, for example, red and orange is difficult.

The triple receptor characteristic of the eye is extremely fortunate as it means that we can generate a range of colours by adding together light sources having just three different wavelengths in various proportions. This process is known as *additive colour matching* which should be clearly distinguished from the subtractive colour matching that occurs with paints and inks. Subtractive matching begins with white light and selectively removes parts of the spectrum by filtering. Additive matching uses coloured light sources which are combined.

5.5   Colour difference signals

An effective colour television system can be made in which only three pure or single wavelength colours or *primaries* can be generated. The primaries need to be similar in wavelength to the peaks of the three receptor responses, but need not be identical. Figure 5.12 shows a rudimentary colour television system. Note that the colour camera is in fact three cameras in one, where each is fitted with a different coloured filter. Three signals, *R*, *G* and *B* must be transmitted to the display which produces three images which must be superimposed to obtain a colour picture.

A monochrome camera produces a single luminance signal *Y* whereas a colour camera produces three signals, or *components*, *R*, *G* and *B* which are essentially monochrome video signals representing an image in each primary colour. *RGB* and *Y* signals are incompatible, yet when colour television was introduced it was a practical necessity that it should be possible to display colour signals on a monochrome display and vice versa.

Creating or *transcoding* a luminance signal from *R*, *G* and *B* is relatively easy. Figure 5.11 showed the spectral response of the eye which has a peak in the green region. Green objects will produce a larger stimulus than red objects of the same brightness, with blue objects producing the least stimulus. A luminance signal can be obtained by adding *R*, *G* and *B* together, not in equal amounts, but in a sum which is *weighted* by the relative response of the eye. Thus:

$$Y = 0.3R + 0.59G + 0.11B$$

If *Y* is derived in this way, a monochrome display will show nearly the same result as if a monochrome camera had been used in the first place.

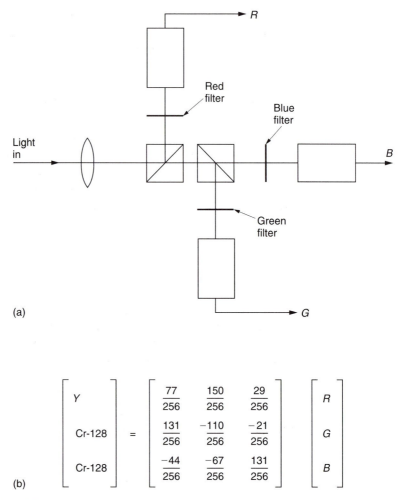

(a)

(b)

$$
\begin{bmatrix} Y \\ Cr\text{-}128 \\ Cr\text{-}128 \end{bmatrix} = \begin{bmatrix} \dfrac{77}{256} & \dfrac{150}{256} & \dfrac{29}{256} \\ \dfrac{131}{256} & \dfrac{-110}{256} & \dfrac{-21}{256} \\ \dfrac{-44}{256} & \dfrac{-67}{256} & \dfrac{131}{256} \end{bmatrix} \begin{bmatrix} R \\ G \\ B \end{bmatrix}
$$

Figure 5.12   (a) A simple colour system uses three primaries and transmits a complete picture for each. This is incompatible with monochrome and uses too much bandwidth. Practical systems use colour difference and luminance signals which are obtained by a weighted calculation as shown in (b).

The results are not identical because of the non-linearities introduced by gamma correction.

As colour pictures require three signals, it should be possible to send Y and two other signals which a colour display could arithmetically convert back to R, G and B. There are two important factors which restrict the form which the other two signals may take. One is to achieve reverse compatibility. If the source is a monochrome camera, it can only produce Y and the other two signals will be completely absent. A colour display should be able to operate on the Y signal only and show a monochrome picture. The other is the requirement to conserve bandwidth for economic reasons.

These requirements are met by sending two *colour difference signals* along with Y. There are three possible colour difference signals, $R - Y$, $B - Y$ and $G - Y$. As the green signal makes the greatest contribution to Y, then the amplitude of $G - Y$ would be the smallest and would be most susceptible to noise. Thus $R - Y$ and $B - Y$ are used in practice as Figure 5.13 shows.

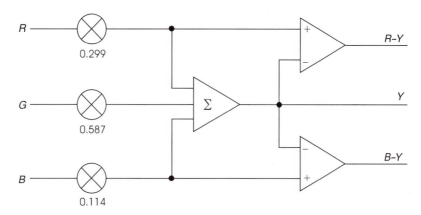

Figure 5.13 Colour components are converted to colour difference signals by the transcoding shown here.

R and B are readily obtained by adding Y to the two colour difference signals. G is obtained by rearranging the expression for Y above such that:

$$G = \frac{Y - 0.3R - 0.11B}{0.59}$$

If a colour CRT is being driven, it is possible to apply inverted luminance to the cathodes and the $R - Y$ and $B - Y$ signals directly to two of the grids so that the tube performs some of the matrixing. It is then only necessary to obtain $G - Y$ for the third grid, using the expression:

$$G - Y = -0.51(R - Y) - 0.186(B - Y)$$

If a monochrome source having only a Y output is supplied to a colour display, $R - Y$ and $B - Y$ will be zero. It is reasonably obvious that if there are no colour difference signals the colour signals cannot be different from one another and $R = G = B$. As a result the colour display can only produce a neutral picture.

The use of colour difference signals is essential for compatibility in both directions between colour and monochrome, but it has a further advantage which follows from the way in which the eye works. In order to produce the highest resolution in the fovea, the eye will use signals from all types of cone, regardless of colour. In order to determine colour the stimuli from three cones must be compared. There is evidence that the nervous system uses some form of colour difference processing to make this possible. As a result the full acuity of the human eye is only available in monochrome. Differences in colour cannot be resolved so well. A further factor is that the lens in the human eye is not achromatic and this means that the ends of the spectrum are not well focused. This is particularly noticeable on blue.

If the eye cannot resolve colour very well there is no point is expending valuable bandwidth sending high-resolution colour signals. Colour difference working allows the luminance to be sent separately at a bit rate which determines the subjective sharpness of the picture. The colour difference signals can be sent with considerably reduced bit rate, as little as one quarter that of luminance, and the human eye is unable to tell.

The overwhelming advantages obtained by using colour difference signals mean that in broadcast and production facilities their use has become almost universal. The technique is equally attractive for compression applications and is retained in MPEG. The outputs from the *RGB* sensors in most cameras are converted directly to Y, $R - Y$ and $B - Y$ in the camera control unit and output in that form. While signals such as Y, R, G and B are *unipolar* or positive only, it should be stressed that colour difference signals are *bipolar* and may meaningfully take on levels below zero volts.

5.6   Progressive or interlaced scan?

Analog video samples in the time domain and vertically down the screen so a two-dimensional vertical/temporal sampling spectrum will result. In a progressively scanned system there is a rectangular matrix of sampling sites vertically and temporally. The rectangular sampling structure of progressive scan is *separable* which means that, for example a vertical manipulation can be performed on frame data without affecting the time axis. The sampling spectrum will be obtained according to section 2.6 and consists of the baseband spectrum repeated as sidebands above and below harmonics of the two dimensional sampling frequencies. The corresponding spectrum is shown in Figure 5.14. The baseband spectrum is in the centre of the diagram, and the repeating sampling sideband spectrum extends vertically and horizontally. The vertical aspects of the star-shaped spectrum result from vertical spatial frequencies in the image. The horizontal aspect is due to image movement. Note that the star shape is rather hypothetical; the actual shape depends heavily on the source material. On a still picture the horizontal dimensions collapse to a line structure. In order to return a progressive scan video signal to a continuous moving picture, a two-dimensional low-pass filter having a rectangular response is required. This is quite feasible as persistence of vision acts as a temporal filter and by sitting far enough away from the screen the finite acuity of the eye acts as a spatial reconstruction filter.

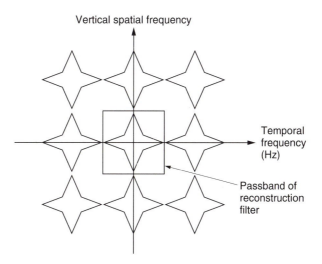

Figure 5.14 The vertical/temporal spectrum of a non-interlaced video signal.

Interlace is actually a primitive form of compression in which the system bandwidth is typically halved by sending only half the frame lines in the first field, with the remaining lines being sent in the second field. The use of interlace has a profound effect on the vertical/temporal spectrum. Figure 5.15 shows that the lowest sampling frequency on the time axis is the frame rate, and the lowest sampling frequency on the vertical axis is the number of lines in a field. The arrangement is called a quincunx pattern because of the similarity to the five of dice. The triangular passband has exactly half the area of the rectangular passband of Figure 5.14 illustrating that half the information rate is available.

As a consequence of the triangular passband of an interlaced signal, if the best vertical frequency response is to be obtained, no motion is allowed. It should be clear from Figure 5.16 that a high vertical spatial frequency resulting from a sharp horizontal edge in the picture is only repeated at frame rate, resulting in an artifact known as *interlace twitter*. Conversely, to obtain the best temporal response, the vertical resolution

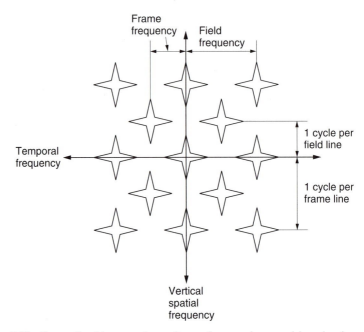

Figure 5.15   The vertical temporal spectrum of monochrome video due to interlace.

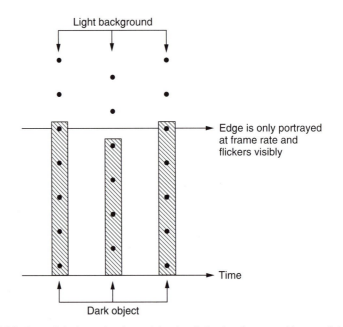

Figure 5.16   In an interlaced system, a horizontal edge is present in one field only and so is presented to the viewer at frame rate not field rate. This is below the critical flicker frequency and is visible as *twitter*.

must be impaired. Thus interlaced systems have poor resolution on moving images, in other words their *dynamic resolution* is poor.

In order to return to a continuous signal, a quincuncial spectrum requires a triangular spatio-temporal low-pass filter. In practice no such filter can be realized. Consequently the sampling sidebands are not filtered out and are visible, particularly the frame rate component. This artifact is visible on an interlaced display even at distances so great that resolution cannot be assessed.

Figure 5.17(a) shows a dynamic resolution analysis of interlaced scanning. When there is no motion, the optic flow axis and the time axis are parallel and the apparent vertical sampling rate is the number of lines in a frame. However, when there is vertical motion, (b), the optic flow axis turns. In the case shown, the sampling structure due to interlace results in the vertical sampling rate falling to one half of its stationary value.

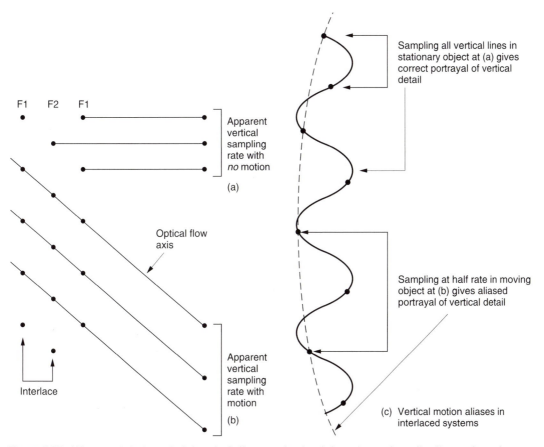

Figure 5.17 When an interlaced picture is stationary, viewing takes place along the time axis as shown in (a). When a vertical component of motion exists, viewing takes place along the optic flow axis. (b) The vertical sampling rate falls to one half its stationary value. (c) The halving in sampling rate causes high spatial frequencies to alias.

Consequently interlace does exactly what would be expected from a half-bandwidth filter. It halves the vertical resolution when any motion with a vertical component occurs. In a practical television system, there is no anti-aliasing filter in the vertical axis and so when the vertical sampling rate of an interlaced system is halved by motion, high spatial frequencies will alias or heterodyne causing annoying artifacts in the picture. This is easily demonstrated.

Figure 5.17(c) shows how a vertical spatial frequency well within the static resolution of the system aliases when motion occurs. In a progressive scan system this effect is absent and the dynamic resolution due to scanning can be the same as the static case.

This analysis also illustrates why interlaced television systems have to have horizontal raster lines. This is because in real life, horizontal motion is more common than vertical. It is easy to calculate the vertical image motion velocity needed to obtain the half-bandwidth speed of interlace, because it amounts to one raster line per field. In 525/60 (NTSC) there are about 500 active lines, so motion as slow as one picture height in 8 seconds will halve the dynamic resolution. In 625/50 (PAL) there are about 600 lines, so the half-bandwidth speed falls to one picture height in 12 seconds. This is why NTSC, with fewer lines and lower bandwidth, doesn't look as soft as it should compared to PAL, because it actually has better dynamic resolution.

The situation deteriorates rapidly if an attempt is made to use interlaced scanning in systems with a lot of lines. In 1250/50, the resolution is halved at a vertical speed of just one picture height in 24 seconds. In other words on real moving video a 1250/50 interlaced system has the same dynamic resolution as a 625/50 progressive system. By the same argument a 1080 I system has the same performance as a 480 P system.

While horizontal raster lines palliate the drawbacks of interlace, they do nothing to help the CRT designer because this arrangement combines the highest scanning frequency with the greatest scanning deflection. With the move to 16:9 aspect ratio, the difficulty becomes even greater. With such a wide tube, it becomes logical to have vertical raster lines so that the deflection of the high-frequency scan (and the current required) is nearly halved. The wide-angle deflection is now only required at the frame rate. The use of interlace prevents this technique.

Interlaced signals are not separable and so processes which are straightforward in progressively scanned systems become more complex in interlaced systems. Compression systems should not be cascaded indiscriminately, especially if they are different. As digital compression techniques based on transforms are now available, it makes no sense to use an interlaced, i.e. compressed, video signal as an input.

Interlaced signals are harder for MPEG-2 to compress.[2] The confusion of temporal and spatial information makes accurate motion estimation more difficult and this reflects in a higher bit rate being required for a given quality. In short, how can a motion estimator accurately measure motion from one field to another when differences between the fields can equally be due to motion, vertical detail or vertical aliasing?

Computer-generated images and film are not interlaced, but consist of discrete frames spaced on a time axis. As digital technology is bringing computers and television closer the use of interlaced transmission is an embarrassing source of incompatibility. The future will bring image-delivery systems based on computer technology and oversampling cameras and displays which can operate at resolutions much closer to the theoretical limits.

Interlace was the best that could be managed with thermionic valve technology sixty years ago, and we should respect the achievement of its developers at a time when things were so much harder. However, we must also recognize that the context in which interlace made sense has disappeared.

5.7 Spatial and temporal redundancy in MPEG

Chapter 1 introduced these concepts in a general sense and now they will be treated with specific reference to MPEG-2. Figure 5.18(a) shows that spatial redundancy is redundancy within a single image, for example repeated pixel values in a large area of blue sky. Temporal redundancy, (b), exists between successive images.

Where temporal compression is used, the current picture is not sent in its entirety; instead the difference between the current picture and the previous picture is sent. The decoder already has the previous picture, and so it can add the difference to make the current picture. A difference picture is created by subtracting every pixel in one picture from the corresponding pixel in another picture. This is trivially easy in a progressively scanned system, but MPEG-2 has had to develop greater complexity so that this can also be done with interlaced pictures. The handling of interlace in MPEG will be detailed later.

A difference picture is an image of a kind, although not a viewable one, and so should contain some kind of spatial redundancy. Figure 5.18(c) shows that MPEG-2 takes advantage of both forms of redundancy. Picture differences are spatially compressed prior to transmission. At the decoder the spatial compression is decoded to recreate the difference picture, then this difference picture is added to the previous picture to complete the decoding process.

Whenever objects move they will be in a different place in successive pictures. This will result in large amounts of difference data. MPEG-2 overcomes the problem using motion compensation. The encoder contains a motion estimator which measures the direction and distance of motion between pictures and outputs these as vectors which are sent to the decoder. When the decoder receives the vectors it uses them to shift data in a previous picture to more closely resemble the current picture. Effectively the vectors are describing the optic flow axis of some moving screen area, along which axis the image is highly redundant. Vectors are bipolar codes which determine the amount of horizontal and vertical shift required.

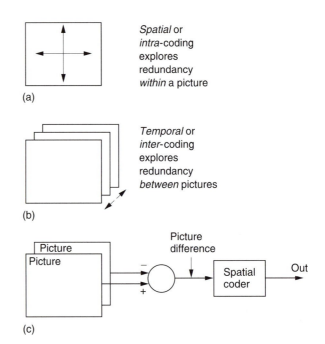

Figure 5.18 (a) Spatial or intra-coding works on individual images. (b) Temporal or inter-coding works on successive images. (c) In MPEG inter-coding is used to create difference images. These are then compressed spatially.

In real images, moving objects do not necessarily maintain their appearance as they move. For example, objects may turn, move into shade or light, or move behind other objects. Consequently motion compensation can never be ideal and it is still necessary to send a picture difference to make up for any shortcomings in the motion compensation.

Figure 5.19 shows how this works. In addition to the motion encoding system, the coder also contains a motion decoder. When the encoder outputs motion vectors, it also uses them locally in the same way that a real decoder will, and is able to produce a *predicted picture* based solely on the previous picture shifted by motion vectors. This is then subtracted from the *actual* current picture to produce a *prediction error* or *residual* which is an image of a kind that can be spatially compressed.

The decoder takes the previous picture, shifts it with the vectors to re-create the predicted picture and then decodes and adds the prediction error to produce the actual picture. Picture data sent as vectors plus prediction error are said to be P coded. The concept of sending a prediction error is a useful approach because it allows both the motion estimation and compensation to be imperfect.

A good motion-compensation system will send just the right amount of vector data. With insufficient vector data, the prediction error will be large, but transmission of excess vector data will also cause the bit rate to rise. There will be an optimum balance which minimizes the sum of the prediction error data and the vector data.

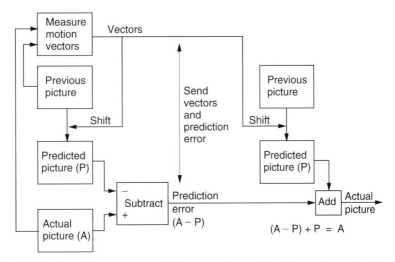

Figure 5.19 A motion-compensated compression system. The coder calculates motion vectors which are transmitted as well as being used locally to create a predicted picture. The difference between the predicted picture and the actual picture is transmitted as a prediction error.

In MPEG-2 the balance is obtained by dividing the screen into areas called *macroblocks* which are 16 luminance pixels square. Each macroblock is steered by a vector. The location of the boundaries of a macroblock are fixed and so the vector does not move the macroblock. Instead the vector tells the decoder where to look in another frame to find pixel data to *fetch* to the macroblock. Figure 5.20(a) shows this concept. The shifting process is generally done by modifying the read address of a RAM using the vector. This can shift by one-pixel steps. MPEG-2 vectors have half-pixel resolution so it is necessary to interpolate between pixels from RAM to obtain half-pixel shifted values.

Real moving objects will not coincide with macroblocks and so the motion compensation will not be ideal but the prediction error makes up

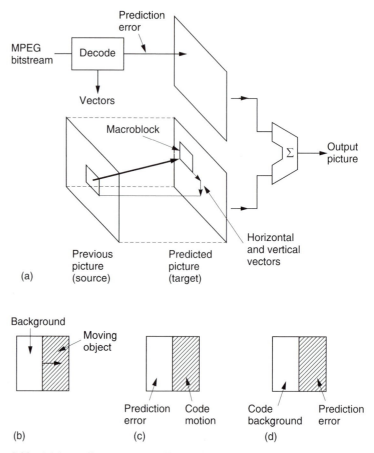

Figure 5.20 (a) In motion compensation, pixel data are brought to a fixed macroblock in the target picture from a variety of places in another picture. (b) Where only part of a macroblock is moving, motion compensation is non-ideal. The motion can be coded (c), causing a prediction error in the background, or the background can be coded (d) causing a prediction error in the moving object.

for any shortcomings. Figure 5.20(b) shows the case where the boundary of a moving object bisects a macroblock. If the system measures the moving part of the macroblock and sends a vector, the decoder will shift the entire block making the stationary part wrong. If no vector is sent, the moving part will be wrong. Both approaches are legal in MPEG-2 because the prediction error sorts out the incorrect values. An intelligent coder might try both approaches to see which required the least prediction error data.

The prediction error concept also allows the use of simple but inaccurate motion estimators in low-cost systems. The greater prediction error data is handled using a higher bit rate. On the other hand, if a precision motion estimator is available, a very high compression factor may be achieved because the prediction error data is minimized. MPEG-2 does not specify how motion is to be measured; it simply defines how a decoder will interpret the vectors. Encoder designers are free to use any motion-estimation system provided that the right vector protocol is created. Chapter 3 contrasted a number of motion-estimation techniques.

Figure 5.21(a) shows that a macroblock contains both luminance and colour difference data at different resolutions. Most of the MPEG-2 Profiles use a 4:2:0 structure which means that the colour is down-sampled by a factor of two in both axes. Thus in a 16 × 16 pixel block, there are only 8 × 8 colour difference sampling sites. MPEG-2 is based upon the 8 × 8 DCT (see section 3.7) and so the 16 × 16 block is the screen area which contains an 8 × 8 colour difference sampling block. Thus in 4:2:0 in each macroblock there are four luminance DCT blocks, one $R - Y$ DCT block and one $B - Y$ DCT block, all steered by the same vector.

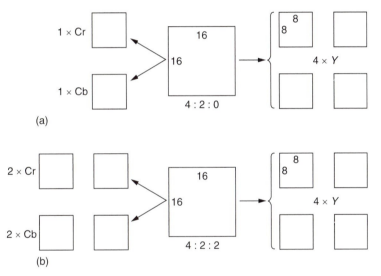

Figure 5.21 The structure of a macroblock. (A macroblock is the screen area steered by *one* vector.) (a) In 4:2:0, there are two chroma DCT blocks per macroblock whereas in 4:2:2 (b) there are four. 4:2:2 needs 33% more data than 4:2:0.

In the 4:2:2 Profile of MPEG-2, shown in Figure 5.21(b), the chroma is not downsampled vertically, and so there is twice as much chroma data in each macroblock which is otherwise substantially the same.

5.8   *I* and *P* coding

Predictive (P) coding cannot be used indefinitely, as it is prone to error propagation. A further problem is that it becomes impossible to decode the transmission if reception begins part-way through. In real video signals, cuts or edits can be present across which there is little redundancy and which make motion estimators throw up their hands.

In the absence of redundancy over a cut, there is nothing to be done but to send the new picture information in absolute form. This is called *I* coding where *I* is an abbreviation of *intra* coding. As *I* coding needs no previous picture for decoding, then decoding can begin at *I* coded information.

MPEG-2 is effectively a toolkit and there is no compulsion to use all the tools available. Thus an encoder may choose whether to use *I* or *P* coding, either once and for all or dynamically on a macroblock by macroblock basis.

For practical reasons, an entire frame may be encoded as *I* macroblocks periodically. This creates a place where the bitstream might be edited or where decoding could begin.

Figure 5.22 shows a typical application of the Simple Profile of MPEG-2. Periodically an *I* picture is created. Between *I* pictures are *P* pictures which are based on the picture before. These *P* pictures predominantly contain macroblocks having vectors and prediction errors. However, it is perfectly legal for *P* pictures to contain *I* macroblocks. This might be useful where, for example, a camera pan introduces new material at the edge of the screen which cannot be created from an earlier picture.

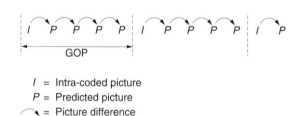

I = Intra-coded picture
P = Predicted picture
⌒ = Picture difference
(vectors plus prediction error)

Figure 5.22 A Simple Profile MPEG-2 signal may contain periodic *I* pictures with a number of *P* pictures between.

Note that although what is sent is called a *P* picture, it is not a picture at all. It is a set of instructions to convert the previous picture into the current picture. If the previous picture is lost, decoding is impossible. An *I* picture together with all of the pictures before the next *I* picture form a *Group of Pictures* (GOP).

5.9   Bidirectional coding

Motion-compensated predictive coding is a useful compression technique, but it does have the drawback that it can only take data from a previous picture. Where moving objects reveal a background this is completely unknown in previous pictures and forward prediction fails.

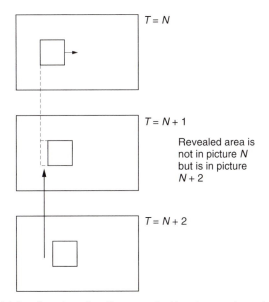

Figure 5.23 In bidirectional coding the revealed background can be efficiently coded by bringing data back from a future picture.

However, more of the background is visible in later pictures. Figure 5.23 shows the concept. In the centre of the diagram, a moving object has revealed some background. The previous picture can contribute nothing, whereas the next picture contains all that is required.

Bidirectional coding is shown in Figure 5.24. A bidirectional or *B* macroblock can be created using a combination of motion compensation

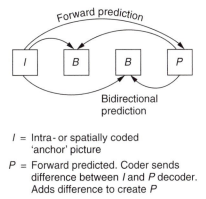

Figure 5.24 In bidirectional coding, a number of *B* pictures can be inserted between periodic forward predicted pictures. See text.

and the addition of a prediction error. This can be done by forward prediction from a previous picture or backward prediction from a subsequent picture. It is also possible to use an average of both forward and backward prediction. On noisy material this may result in some reduction in bit rate. The technique is also a useful way of portraying a dissolve.

The averaging process in MPEG-2 is a simple linear interpolation which works well when only one *B* picture exists between the reference pictures before and after. A larger number of *B* pictures would require weighted interpolation but MPEG-2 does not support this.

Typically two *B* pictures are inserted between *P* pictures or between *I* and *P* pictures. As can be seen, *B* pictures are never predicted from one another, only from *I* or *P* pictures. A typical GOP for broadcasting purposes might have the structure *IBBPBBPBBPBB*. Note that the last *B* pictures in the GOP require the *I* picture in the next GOP for decoding and so the GOPs are not truly independent. Independence can be obtained by creating a *closed GOP* which may contain *B* pictures but which ends with a *P* picture. It is also legal to have a *B* picture in which every macroblock is forward predicted, needing no future picture for decoding.

Bidirectional coding is very powerful. Figure 5.25 is a constant quality curve showing how the bit rate changes with the type of coding. On the left, only *I* or spatial coding is used, whereas on the right an *IBBP* structure is used. This means that there are two bidirectionally coded pictures in between a spatially coded picture (*I*) and a forward predicted picture (*P*). Note how for the same quality the system which only uses spatial coding needs two and a half times the bit rate that the bidirectionally coded system needs.

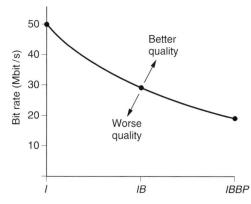

Figure 5.25 Bidirectional coding is very powerful as it allows the same quality with only 40% of the bit rate of intra-coding. However, the encoding and decoding delays must increase. Coding over a longer time span is more efficient but editing is more difficult.

Clearly information in the future has yet to be transmitted and so is not normally available to the decoder. MPEG-2 gets around the problem by sending pictures in the wrong order. Picture reordering requires delay in the encoder and a delay in the decoder to put the order right again. Thus the overall codec delay must rise when bidirectional coding is used. This is quite consistent with Figure 1.5 which showed that as the compression factor rises the latency must also rise.

Figure 5.26 shows that although the original picture sequence is *IBBPBBPBBIBB* . . ., this is transmitted as *IPBBPBBIBB* . . . so that the future picture is already in the decoder before bidirectional decoding begins. Note that the *I* picture of the next GOP is actually sent before the last *B* pictures of the current GOP.

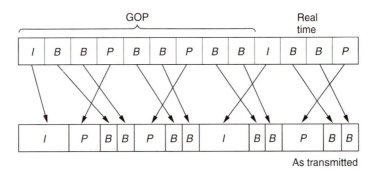

Figure 5.26 Comparison of pictures before and after compression showing sequence change and varying amount of data needed by each picture type. *I, P, B* pictures use unequal amounts of data.

Figure 5.26 also shows that the amount of data required by each picture is dramatically different. *I* pictures have only spatial redundancy and so need a lot of data to describe them. *P* pictures need less data because they are created by shifting the *I* picture with vectors and then adding a prediction error picture. *B* pictures need the least data of all because they can be created from *I* or *P*.

With pictures requiring a variable length of time to transmit, arriving in the wrong order, the decoder needs some help. This takes the form of picture-type flags and time stamps which will be described in section 6.2.

5.10   Coding applications

Figure 5.27 shows a variety of GOP structures. The simplest is the *III* . . . sequence in which every picture is intra-coded. Pictures can be fully decoded without reference to any other pictures and so editing is straightforward. However, this approach requires about two-and-one-half times the bit rate of a full bidirectional system. Bidirectional coding is most useful for final delivery of post-produced material either by broadcast or on prerecorded media as there is then no editing requirement. As a compromise the *IBIB* . . . structure can be used which has some of the bit rate advantage of bidirectional coding but without too much latency. It is possible to edit an *IBIB* stream by performing some processing. If it is required to remove the video following a *B* picture, that *B* picture could not be decoded because it needs *I* pictures either side of

I I I I I I...	*I* only freely editable, needs high bit rate
I P P P P I P...	Forward predicted only, needs less decoder memory, used in Simple Profile
I B B P B B P B...	Forward and bidirectional, best compression factor, needs large decoder memory, hard to edit
I B I B I B...	Lower bit rate than *I* only, editable with moderate processing

Figure 5.27 Various possible GOP structures used with MPEG. See text for details.

it for bidirectional decoding. The solution is to decode the *B* picture first, and then re-encode it with forward prediction only from the previous *I* picture. The subsequent *I* picture can then be replaced by an edit process. Some quality loss is inevitable in this process but this is acceptable in applications such as ENG and industrial video.

5.11   Spatial compression

Spatial compression in MPEG-2 is used in *I* pictures on actual picture data and in *P* and *B* pictures on prediction error data. MPEG-2 uses the Discrete Cosine Transform described in section 3.7. The DCT works on blocks and in MPEG-2 these are 8×8 pixels. Section 5.7 showed how the macroblocks of the motion-compensation structure are designed so they can be broken down into 8×8 DCT blocks. In a 4:2:0 macroblock there will be six DCT blocks whereas in a 4:2:2 macroblock there will be eight.

Figure 5.28 shows the table of basis functions or *wave table* for an 8×8 DCT. Adding these two-dimensional waveforms together in different proportions will give any original 8×8 pixel block. The coefficients of the DCT simply control the proportion of each wave which is added in the inverse transform. The top-left wave has no modulation at all because it

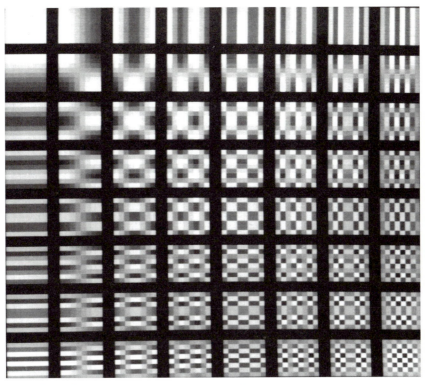

Figure 5.28 The discrete cosine transform breaks up an image area into discrete frequencies in two dimensions. The lowest frequency can be seen here at the top-left corner. Horizontal frequency increases to the right and vertical frequency increases downwards.

conveys the DC component of the block. This coefficient will be a unipolar (positive only) value in the case of luminance and will typically be the largest value in the block as the spectrum of typical video signals is dominated by the DC component.

Increasing the DC coefficient adds a constant amount to every pixel. Moving to the right the coefficients represent increasing horizontal spatial frequencies and moving downwards the coefficients represent increasing vertical spatial frequencies. The bottom-right coefficient represents the highest diagonal frequencies in the block. All these coefficients are bipolar, where the polarity indicates whether the original spatial waveform at that frequency was inverted.

Figure 5.29 shows a one-dimensional example of an inverse transform. The DC coefficient produces a constant level throughout the pixel block. The remaining waves in the table are AC coefficients. A zero coefficient would result in no modulation, leaving the DC level unchanged. The wave next to the DC component represents the lowest frequency in the transform which is half a cycle per block. A positive coefficient would make the left side of the block brighter and the right side darker whereas a negative coefficient would do the opposite. The magnitude of the coefficient determines the amplitude of the wave which is added. Figure 5.29 also shows that the next wave has a frequency of one cycle per block, i.e. the block is made brighter at both sides and darker in the middle.

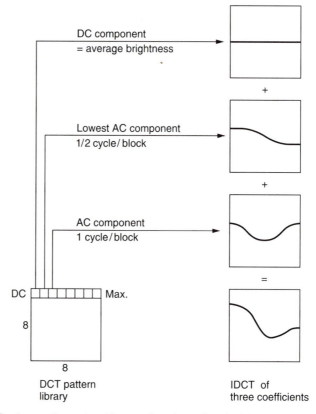

Figure 5.29 A one-dimensional inverse transform. See text for details.

Consequently an inverse DCT is no more than a process of mixing various pixel patterns from the wave table where the relative amplitudes and polarity of these patterns are controlled by the coefficients. The original transform is simply a mechanism which finds the coefficient amplitudes from the original pixel block.

The DCT itself achieves no compression at all. Sixty-four pixels are converted to sixty-four coefficients. However, in typical pictures, not all coefficients will have significant values; there will often be a few dominant coefficients. The coefficients representing the higher two-dimensional spatial frequencies will often be zero or of small value in large areas, due to blurring or simply plain undetailed areas before the camera.

Statistically, the further from the top-left corner of the wave table the coefficient is, the smaller will be its magnitude. Coding gain (the technical term for reduction in the number of bits needed) is achieved by transmitting the low-valued coefficients with shorter wordlengths. The zero-valued coefficients need not be transmitted at all. Thus it is not the DCT which compresses the data, it is the subsequent processing. The DCT simply expresses the data in a form which makes the subsequent processing easier.

Higher compression factors require the coefficient wordlength to be further reduced using requantizing. Coefficients are divided by some factor which increases the size of the quantizing step. The smaller number of steps which results permits coding with fewer bits, but, of course, with an increased quantizing error. The coefficients will be multiplied by a reciprocal factor in the decoder to return to the correct magnitude.

Inverse transforming a requantized coefficient means that the frequency it represents is reproduced in the output with the wrong amplitude. The difference between original and reconstructed amplitude is regarded as a noise added to the wanted data. Figure 5.10 showed that the visibility of such noise is far from uniform. The maximum sensitivity is found at DC and falls thereafter. As a result the top-left coefficient is often treated as a special case and left unchanged. It may warrant more error protection than other coefficients.

MPEG-2 takes advantage of the falling sensitivity to noise. Prior to requantizing, each coefficient is divided by a different weighting constant as a function of its frequency. Figure 5.30 shows a typical weighting

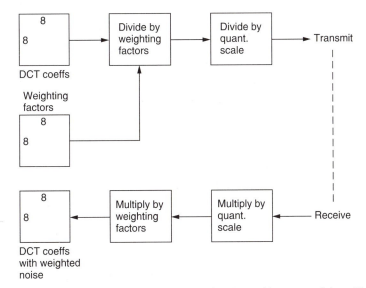

Figure 5.30 Weighting is used to make the noise caused by requantizing different at each frequency.

process. Naturally the decoder must have a corresponding inverse weighting. This weighting process has the effect of reducing the magnitude of high-frequency coefficients disproportionately. Clearly different weighting will be needed for colour difference data as colour is perceived differently.

P and *B* pictures are decoded by adding a prediction error image to a reference image. That reference image will contain weighted noise. One purpose of the prediction error is to cancel that noise to prevent tolerance build-up. If the prediction error were also to contain weighted noise this result would not be obtained. Consequently prediction error coefficients are flat weighted.

When forward prediction fails, such as in the case of new material introduced in a *P* picture by a pan, *P* coding would set the vectors to zero and encode the new data entirely as an unweighted prediction error. In this case it is better to encode that material as an *I* macroblock because then weighting can be used and this will require fewer bits.

Requantizing increases the step size of the coefficients, but the inverse weighting in the decoder results in step sizes which increase with frequency. The larger step size increases the quantizing noise at high frequencies where it is less visible. Effectively the noise floor is shaped to match the sensitivity of the eye. The quantizing table in use at the encoder can be transmitted to the decoder periodically in the bitstream.

5.12   Scanning and run-length/variable-length coding

Study of the signal statistics gained from extensive analysis of real material is used to measure the probability of a given coefficient having a given value. This probability turns out to be highly non-uniform, suggesting the possibility of a variable-length encoding for the coefficient values. On average, the higher the spatial frequency, the lower the value of a coefficient will be. This means that the value of a coefficient falls as a function of its radius from the DC coefficient.

Typical material often has many coefficients which are zero valued, especially after requantizing. The distribution of these also follows a pattern. The non-zero values tend to be found in the top-left-hand corner of the DCT block, but as the radius increases, not only do the coefficient values fall, but it becomes increasingly likely that these small coefficients will be interspersed with zero-valued coefficients. As the radius increases further it is probable that a region where all coefficients are zero will be entered.

MPEG-2 uses all these attributes of DCT coefficients when encoding a coefficient block. By sending the coefficients in an optimum order, by describing their values with Huffman coding and by using run-length encoding for the zero-valued coefficients it is possible to achieve a significant reduction in coefficient data which remains entirely lossless. Despite the complexity of this process, it does contibute to improved picture quality because for a given bit rate lossless coding of the coefficients must be better than requantizing, which is lossy. Of course, for lower bit rates both will be required.

It is an advantage to scan in a sequence where the largest coefficient values are scanned first. Then the next coefficient is more likely to be zero than the previous one. With progressively scanned material, a regular zig-zag scan begins in the top-left corner and ends in the bottom-right corner as shown in Figure 5.31. Zig-zag scanning means that significant values are more likely to be transmitted first, followed by the zero values. Instead of coding these zeros, an unique 'end of block' (EOB) symbol is transmitted instead.

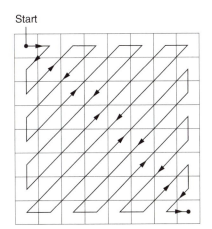

Figure 5.31   The zig-zag scan for a progressively scanned image.

As the zig-zag scan approaches the last finite coefficient it is increasingly likely that some zero value coefficients will be scanned. Instead of transmitting the coefficients as zeros, the *zero-run-length*, i.e. the number of zero valued coefficients in the scan sequence, is encoded into the next non-zero coefficient which is itself variable-length coded. This combination of run-length and variable-length coding is known as RLC/ VLC in MPEG-2.

The DC coefficient is handled separately because it is differentially coded and this discussion relates to the AC coefficients. Three items need to be handled for each coefficient: the zero-run-length prior to this coefficient, the wordlength and the coefficient value itself. The word length needs to be known by the decoder so that it can correctly parse the bitstream. The wordlength of the coefficient is expressed directly as an integer called the *size*.

Figure 5.32(a) shows that a two-dimensional run/size table is created. One dimension expresses the zero-run-length; the other the size. A run length of zero is obtained when adjacent coefficients are non-zero, but a code of 0/0 has no meaningful run/size interpretation and so this bit pattern is used for the end-of-block (EOB) symbol.

In the case where the zero-run-length exceeds 14, a code of 15/0 is used signifying that there are fifteen zero-valued coefficients. This is then followed by another run/size parameter whose run-length value is added to the previous fifteen.

The run/size parameters contain redundancy because some combinations are more common than others. Figure 5.32(b) shows that each run/size value is converted to a variable-length Huffman codeword for transmission. As was shown in section 1.5, the Huffman codes are

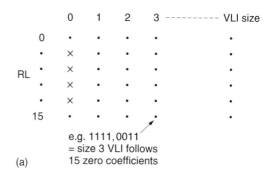

(a)

Size Run	0	1	2	3 → etc.
0	1010 (EOB)	00	01	100
1	–	1100	11011	
2	–	11100	11111001	
3	–	111010	111110111	
4	–	111011	1111111000	
5	–	1111010		

etc.

(b)

Figure 5.32 Run-length and variable-length coding simultaneously compresses runs of zero-valued coefficients and describes the wordlength of a non-zero coefficient.

designed so that short codes are never a prefix of long codes so that the decoder can deduce the parsing by testing an increasing number of bits until a match with the look-up table is found. Having parsed and decoded the Huffman run/size code, the decoder then knows what the coefficient wordlength will be and can correctly parse that.

The variable-length coefficient code has to describe a bipolar coefficient, i.e one which can be positive or negative. Figure 5.32(c) shows that for a particular size, the coding scale has a certain gap in it. For example, all values from –7 to +7 can be sent by a size 3 code, so a size 4 code only has to send the values of –15 to –8 and +8 to +15. The coefficient code is sent as a pure binary number whose value ranges from all zeros to all ones where the maximum value is a function of the size. The number

etc.

Coefficient to be transmitted	Coefficient code	Size parameter
15	1 1 1 1	
14	1 1 1 0	
13	1 1 0 1	
12	1 1 0 0	4
11	1 0 1 1	
10	1 0 1 0	
9	1 0 0 1	
8	1 0 0 0	
7	1 1 1	
6	1 1 0	3
5	1 0 1	
4	1 0 0	
3	1 1	2
2	1 0	
1	1	1
– 1	0	
– 2	0 1	2
– 3	0 0	
– 4	0 1 1	
– 5	0 1 0	3
– 6	0 0 1	
– 7	0 0 0	
– 8	0 1 1 1	
– 9	0 1 1 0	
– 10	0 1 0 1	
– 11	0 1 0 0	4
– 12	0 0 1 1	
– 13	0 0 1 0	
– 14	0 0 0 1	
– 15	0 0 0 0	

(c)

etc.

Figure 5.32 (c)

range is divided into two, the lower half of the codes specifying negative values and the upper half specifying positive.

In the case of positive numbers, the transmitted binary value is the actual coefficient value, whereas in the case of negative numbers a constant must be subtracted which is a function of the size. In the case of a size 4 code, the constant is 15_{10}. Thus a size 4 parameter of 0111_2 (7_{10}) would be interpreted as $7 - 15 = -8$. A size of 5 has a constant of 31 so a transmitted coded of 01010_2 (10_2) would be interpreted as $10 - 31 = -21$.

This technique saves a bit because, for example, 63 values from -31 to $+31$ are coded with only 5 bits having only 32 combinations. This is possible because that extra bit is effectively encoded into the run/size parameter.

Figure 5.33 shows the whole spatial coding subsystem. Macroblocks are subdivided into DCT blocks and the DCT is calculated. The resulting coefficients are multiplied by the weighting matrix and then requantized. The coefficients are then reordered by the zig-zag scan so that full advantage can be taken of run-length and variable-length coding. The last non-zero coefficient in the scan is followed by the EOB symbol.

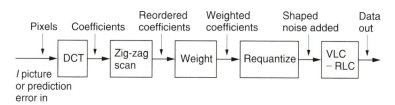
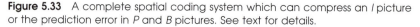

Figure 5.33   A complete spatial coding system which can compress an *I* picture or the prediction error in *P* and *B* pictures. See text for details.

In predictive coding, sometimes the motion-compensated prediction is nearly exact and so the prediction error will be almost zero. This can also happen on still parts of the scene. MPEG-2 takes advantage of this by sending a code to tell the decoder there is no prediction error data for the macroblock concerned.

The success of temporal coding depends on the accuracy of the vectors. Trying to reduce the bit rate by reducing the accuracy of the vectors is false economy as this simply increases the prediction error. Consequently for a given GOP structure it is only in the the spatial coding that the overall bit rate is determined. The RLC/VLC coding is lossless and so its contribution to the compression cannot be varied. If the bit rate is too high, the only option is to increase the size of the coefficient-requantizing steps. This has the effect of shortening the wordlength of large coefficients, and rounding small coefficients to zero, so that the bit rate goes down. Clearly if taken too far the picture quality will also suffer because at some point the noise floor will become visible as some form of artifact.

5.13   A bidirectional coder

MPEG-2 does not specify how an encoder is to be built or what coding decisions it should make. Instead it specifies the protocol of the bitstream at the output. As a result the coder shown in Figure 5.34 is only an example.

Figure 5.34(a) shows the component parts of the coder. At the input is a chain of picture stores which can be bypassed for reordering purposes. This allows a picture to be encoded ahead of its normal timing when bidirectional coding is employed.

At the centre is a dual-motion estimator which can simultaneously measure motion between the input picture, an earlier picture and a later picture. These reference pictures are held in frame stores. The vectors from the motion estimator are used locally to shift a picture in a frame store to form a predicted picture. This is subtracted from the input picture to produce a prediction error picture which is then spatially coded.

The bidirectional encoding process will now be described. A GOP begins with an *I* picture which is intra-coded. In Figure 5.34(b) the *I* picture emerges from the reordering delay. No prediction is possible on an *I* picture so the motion estimator is inactive. There is no predicted picture and so the prediction error subtractor is set simply to pass the input. The only processing which is active is the forward spatial coder which describes the picture with DCT coefficients. The output of the forward spatial coder is locally decoded and stored in the past picture frame store.

The reason for the spatial encode/decode is that the past picture frame store now contains exactly what the decoder frame store will contain, including the effects of any requantizing errors. When the same picture is used as a reference at both ends of a differential coding system, the errors will cancel out.

Having encoded the *I* picture, attention turns to the *P* picture. The input sequence is *IBBP,* but the transmitted sequence must be *IPBB.* Figure 5.34(c) shows that the reordering delay is bypassed to select the *P* picture. This passes to the motion estimator which compares it with the *I* picture and outputs a vector for each macroblock. The forward predictor uses these vectors to shift the *I* picture so that it more closely resembles the *P* picture. The predicted picture is then subtracted from the actual picture to produce a forward prediction error. This is then spatially coded. Thus the *P* picture is transmitted as a set of vectors and a prediction error image.

The *P* picture is locally decoded in the right-hand decoder. This takes the forward predicted picture and adds the decoded prediction error to obtain exactly what the decoder will obtain.

Figure 5.34(d) shows that the encoder now contains a *I* picture in the left store and a *P* picture in the right store. The reordering delay is

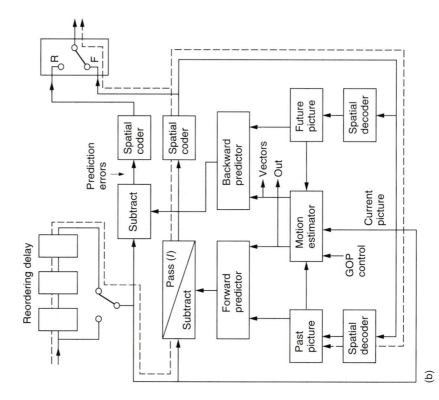

(a)

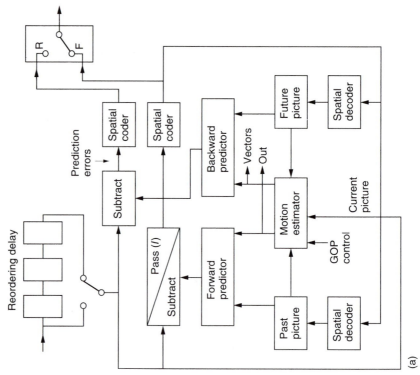

(b)

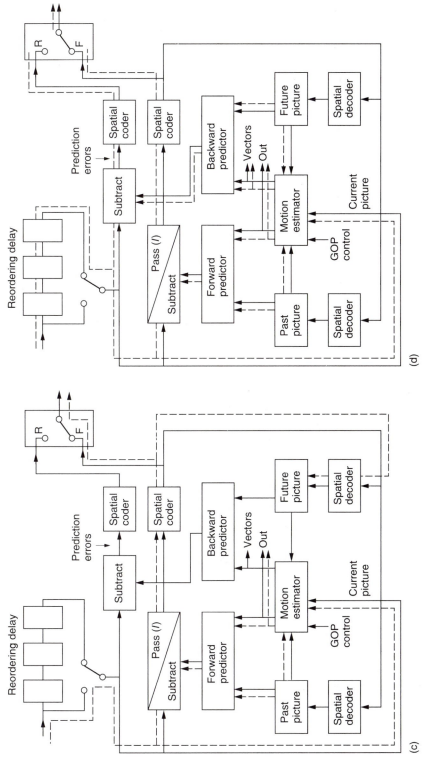

Figure 5.34 A bidirectional coder. (a) The essential components. (b) Signal flow when coding an *I* picture. (c) Signal flow when coding a *P* picture. (d) Signal flow when bidirectional coding.

(c)

(d)

reselected so that the first *B* picture can be input. This passes to the motion estimator where it is compared with both the *I* and *P* pictures to produce forward and backward vectors. The forward vectors go to the forward predictor to make a *B* prediction from the *I* picture. The backward vectors go to the backward predictor to make a *B* prediction from the *P* picture. These predictions are simultaneously subtracted from the actual *B* picture to produce a forward prediction eror and a backward prediction error. These are then spatially encoded. The encoder can then decide which direction of coding resulted in the best prediction; i.e. the smallest prediction error.

Not shown in the interests of clarity is a third signal path which creates a predicted *B* picture from the average of forward and backward predictions. This is subtracted from the input picture to produce a third prediction error. In some circumstances this prediction error may use less data than either forward of backward prediction alone.

As *B* pictures are never used to create other pictures, the decoder does not locally decode the *B* picture. After decoding and displaying the *B* picture the decoder will discard it. At the encoder the *I* and *P* pictures remain in their frame stores and the second *B* picture is input from the reordering delay.

Following the encoding of the second *B* picture, the encoder must reorder again to encode the second *P* picture in the GOP. This will be locally decoded and will replace the *I* picture in the left store. The stores and predictors switch designation because the left store is now a future *P* picture and the right store is now a past *P* picture. *B* pictures between them are encoded as before.

5.14   Slices

There is still some redundancy in the output of a bidirectional coder and MPEG-2 is remarkably diligent in finding it. In *I* pictures, the DC coefficient describes the average brightness of an entire DCT block. In real video the DC component of adjacent blocks will be similar much of the time. A saving in bit rate can be obtained by differentially coding the DC coefficient.

In *P* and *B* pictures this is not done because these are prediction errors not actual images and the statistics are different. However, *P* and *B* pictures send vectors and instead the redundancy in these is explored. In a large moving object, many macroblocks will be moving at the same velocity and their vectors will be the same. Thus differential vector coding will be advantageous.

As has been seen above, differential coding cannot be used indiscriminately as it is prone to error propagation. Periodically absolute DC coefficients and vectors must be sent and the *slice* is the logical structure which supports this mechanism. In *I* pictures, the first DC coefficient in a slice is sent in absolute form, whereas the subsequent coefficients are sent differentially. In *P* or *B* pictures, the first vector in a slice is sent in absolute form, but the subsequent vectors are differential.

Slices are horizontal picture strips which are one macroblock (16 pixels) high and which proceed from left to right across the screen. The sides of the picture must coincide with the beginning or the end of a slice in DVB, but otherwise the encoder is free to decide how big slices should be and where they begin.

In the case of a central dark building silhouetted against the bright sky, there would be two large changes in the DC coefficients, one at each edge of the building. It may be advantageous to the encoder to break the width of the picture into three slices, one each for the left and right areas of sky and one for the building. In the case of a large moving object, different slices may be used for the object and the background.

Each slice contains its own synchronizing pattern, so following a transmission error, correct decoding can resume at the next slice. Slice size can also be matched to the characteristics of the transmission channel. For example, in an error-free transmission system the use of a large number of slices in a packet simply wastes data capacity on surplus synchronizing patterns. However, in a non-ideal system it might be advantageous to have frequent resynchronizing.

5.15   Handling interlaced pictures

Spatial coding, predictive coding and motion compensation can still be performed using interlaced source material at the cost of considerable complexity. Despite that complexity, MPEG-2 cannot be expected to perform as well with interlaced material.

Figure 5.35 shows that in an incoming interlaced frame there are two fields each of which contain half of the lines in the frame. In MPEG-2 these are known as the *top field* and the *bottom field*. In video from a camera, these fields represent the state of the image at two different times. Where there is little image motion, this is unimportant and the fields can be combined obtaining more effective compression. However, in the presence of motion the fields become increasingly decorrelated because of the displacement of moving objects from one field to the next.

This characteristic determines that MPEG-2 must be able to handle fields independently or together. This dual approach permeates all aspects of MPEG-2 and affects the definition of pictures, macroblocks, DCT blocks and zig-zag scanning.

Figure 5.35 also shows how MPEG-2 designates interlaced fields. In picture types *I*, *P* and *B*, the two fields can be superimposed to make a *frame-picture* or the two fields can be coded independently as two *field-pictures*. As a third possibility, in *I* pictures only, the bottom field-picture can be predictively coded from the top field-picture to make an *IP* frame-picture.

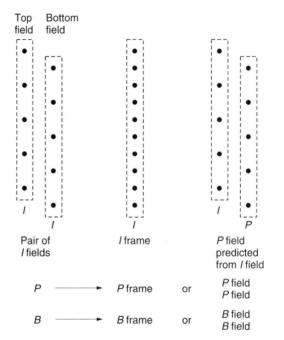

Figure 5.35 An interlaced frame consists of top and bottom fields. MPEG-2 can code a frame in the ways shown here.

A frame-picture is one in which the macroblocks contain lines from both field types over a picture area 16 scan lines high. Each luminance macroblock contains the usual four DCT blocks but there are two ways in which these can be assembled. Figure 5.36(a) shows how a frame is divided into *frame DCT* blocks. This is identical to the progressive scan approach in that each DCT block contains eight contiguous picture lines. In 4:2:0, the colour difference signals have been downsampled by a factor of two and shifted as was shown in Figure 2.17. Figure 5.36(a) also shows how one 4:2:0 DCT block contains the chroma data from 16 lines in two fields.

Even small amounts of motion in any direction can destroy the correlation between odd and even lines and a frame DCT will result in an excessive number of coefficients. Figure 5.36(b) shows that instead the luminance component of a frame can also be divided into *field DCT*

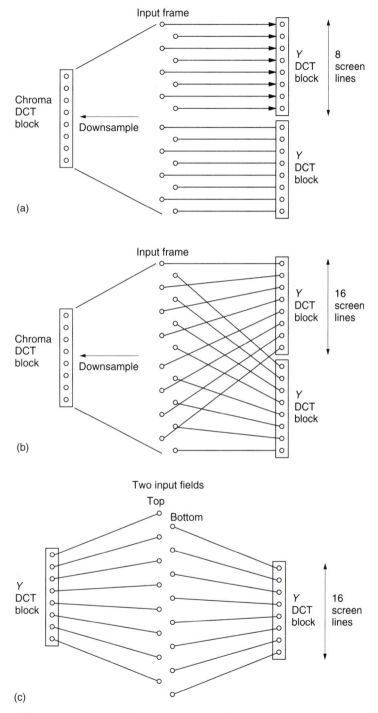

Figure 5.36 (a) In Frame-DCT, a picture is effectively de-interlaced. (b) In Field-DCT, each DCT block only contains lines from one field, but over twice the screen area. (c) The same DCT content results when field-pictures are assembled into blocks.

blocks. In this case one DCT block contains odd lines and the other contains even lines. In this mode the chroma still produces one DCT block from both fields as in Figure 5.36(a).

When an input frame is designated as two *field-pictures*, the macroblocks come from a screen area which is 32 lines high. Figure 5.36(c) shows that the DCT blocks contain the same data as if the input frame had been designated a *frame-picture* but with *field DCT*. Consequently it is only frame-pictures which have the option of field or frame DCT. These may be selected by the encoder on a macroblock-by-macroblock basis and, of course, the resultant bitstream must specify what has been done.

In a frame which contains a small moving area, it may be advantageous to encode as a frame-picture with frame DCT except in the moving area where field DCT is used. This approach may result in fewer bits than coding as two field-pictures.

In a field-picture and in a frame-picture using field DCT, a DCT block contains lines from one field type only and this must have come from a screen area 16 scan lines high, whereas in progressive scan and frame DCT the area is only 8 scan lines high. A given vertical spatial frequency in the image is sampled at points twice as far apart which is interpreted by the field DCT as a doubled spatial frequency, whereas there is no change in the horizontal spectrum.

Following the DCT calculation, the coefficient distribution will be different in field-pictures and field DCT frame-pictures. In these cases, the probability of coefficients is not a constant function of radius from the DC coefficient as it is in progressive scan, but is elliptical where the ellipse is twice as high as it is wide.

Using the standard 45° zig-zag scan with this different coefficient distribution would not have the required effect of putting all the significant coefficients at the beginning of the scan. To achieve this requires a different zig-zag scan, which is shown in Figure 5.37. This scan, sometimes known as the Yeltsin walk, attempts to match the elliptical probability of interlaced coefficients with a scan slanted at 67.5° to the vertical. This is clearly suboptimal, and is one of the reasons why MPEG-2 does not work so well with interlaced video.

Motion estimation is more difficult in an interlaced system. Vertical detail can result in differences between fields and this reduces the quality of the match. Fields are vertically subsampled without filtering and so contain alias products. This aliasing will mean that the vertical waveform representing a moving object will not be the same in successive pictures and this will also reduce the quality of the match.

Even when the correct vector has been found, the match may be poor so the estimator fails to recognize it. If it is recognized, a poor match means that the quality of the prediction in *P* and *B* pictures will be poor and so a large prediction error or residual has to be transmitted. In an attempt to reduce the residual, MPEG-2 allows field-pictures to use motion-compensated prediction from either the adjacent field or from the same field type in another frame. In this case the encoder will use the better match. This technique can also be used in areas of frame-pictures which use field DCT.

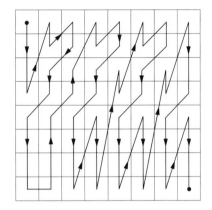

Figure 5.37 The zig-zag scan for an interlaced image has to favour vertical frequencies twice as much as horizontal.

The motion compensation of MPEG-2 has half-pixel resolution and this is inherently compatible with interlace because an interpolator must be present to handle the half-pixel shifts. Figure 5.38(a) shows that in an interlaced system, each field contains half of the frame lines and so interpolating half-way between lines of one field type will actually create values lying on the sampling structure of the other field type. Thus it is equally possible for a predictive system to decode a given field type based on pixel data from the other field type or of the same type.

If when using predictive coding from the other field type the vertical motion vector contains a half-pixel component, then no interpolation is needed because the act of transferring pixels from one field to another results in such a shift.

Figure 5.38(b) shows that a macroblock in a given P field-picture can be encoded using a vector which shifts data from the previous field or from the field before that, irrespective of which frames these fields occupy. As noted above, field-picture macroblocks come from an area of screen 32 lines high and in this means that the vector density is halved resulting in larger prediction errors at the boundaries of moving objects.

As an option, field-pictures can restore the vector density by using 16 $\times$ 8 motion compensation where separate vectors are used for the top and bottom halves of the macroblock. Frame-pictures can also use 16 $\times$ 8

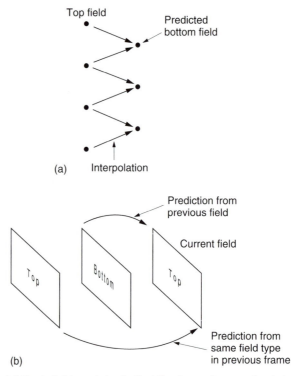

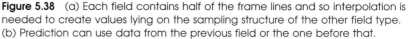

Figure 5.38 (a) Each field contains half of the frame lines and so interpolation is needed to create values lying on the sampling structure of the other field type. (b) Prediction can use data from the previous field or the one before that.

motion compensation in conjunction with field DCT. While the 2×2 DCT block luminance structure of a macroblock can easily be divided vertically in two, in 4:2:0 the same screen area is represented by only one chroma macroblock of each component type. As it cannot be divided in half, this chroma is deemed to belong to the luminance DCT blocks of the upper field. In 4:2:2 no such difficulty arises.

MPEG-2 supports interlace simply because interlaced video exists in legacy systems and there is a requirement to compress it. However, where the opportunity arises to define a new system, interlace should be avoided. Legacy interlaced source material should be handled using a motion-compensated de-interlacer prior to compression in the progressive domain.

5.16 An MPEG-2 coder

Figure 5.39 shows the complete coder. The bidirectional coder outputs coefficients and vectors, and the quantizing table in use. The vectors of P and B pictures and the DC coefficients of I pictures are differentially encoded in slices and the remaining coefficients are RLC/VLC coded. The multiplexer assembles all of this data into a single bitstream called an Elementary Stream. The output of the encoder is a buffer which absorbs the variations in bit rate between different picture types. The buffer output has a constant bit rate determined by the demand clock. This comes from the transmission channel or storage device. If the bit rate is low, the buffer will tend to fill up, whereas if it is high the buffer will tend to empty. The buffer content is used to control the severity of the requantizing in the spatial coders. The more the buffer fills, the bigger the requantizing steps get.

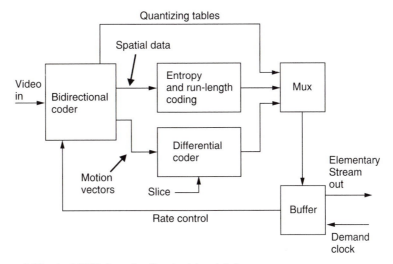

Figure 5.39 An MPEG 2 coder. See text for details.

The buffer in the decoder has a finite capacity and the encoder must model the decoder's buffer occupancy so that it neither overflows nor underflows. An overflow might occur if an *I* picture is transmitted when the buffer content is already high. The buffer occupancy of the decoder depends somewhat on the memory-access strategy of the decoder. Instead of defining a specific buffer size, MPEG-2 defines the size of a particular mathematical model of a hypothetical buffer. The decoder designer can use any strategy which implements the model, and the encoder can use any strategy which doesn't overflow or underflow the model. The Elementary Stream has a parameter called the video buffer verifier (VBV) which defines the minimum buffering assumptions of the encoder.

As was seen in Chapter 1, buffering is one way of ensuring constant quality when picture entropy varies. An intelligent coder may run down the buffer contents in anticipation of a difficult picture sequence so that a large amounts of data can be sent.

MPEG-2 does not define what a decoder should do if a buffer underflow or overflow occurs, but since both irrecoverably lose data it is obvious that there will be more or less of an interruption to the decoding. Even a small loss of data may cause loss of synchronization and in the case of long GOP the lost data may make the rest of the GOP undecodable. A decoder may choose to repeat the last properly decoded picture until it can begin to operate correctly again.

Buffer problems occur if the VBV model is violated. If this happens then more than one underflow or overflow can result from a single violation. Switching an MPEG bitstream can cause a violation because the two encoders concerned may have radically different buffer occupancy at the switch.

5.17   The Elementary Stream

Figure 5.40 shows the structure of the Elementary Stream from an MPEG-2 encoder. The structure begins with a set of coefficients representing a DCT block. Six or eight DCT blocks form the luminance and chroma content of one macroblock. In *P* and *B* pictures a macroblock will be associated with a vector for motion compensation. Macroblocks are associated into slices in which DC coefficients of *I* pictures and vectors in *P* and *B* pictures are differentially coded. An arbitrary number of slices forms a picture and this needs *I/P/B* flags describing the type of picture it is. The picture may also have a global vector which efficiently deals with

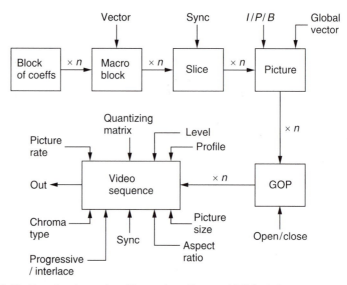

Figure 5.40 The structure of an Elementary Stream. MPEG defines the syntax precisely.

pans. Several pictures form a Group of Pictures (GOP). The GOP begins with an *I* picture and may or may not include *P* and *B* pictures in a structure which may vary dynamically.

Several GOPs form a Sequence which bgins with a Sequence header containing important data to help the decoder. It is possible to repeat the header within a sequence, and this helps lock-up in random access applications. The Sequence header describes the MPEG-2 profile and level, whether the video is progressive or interlaced, whether the chroma is 4:2:0 or 4:2:2, the size of the picture and the aspect ratio of the pixels. The quantizing matrix used in the spatial coder can also be sent. The sequence begins with a standardized bit pattern which is detected by a decoder to synchronize the deserialization.

5.18   An MPEG-2 decoder

The decoder is only defined by implication from the definitions of syntax and any decoder which can correctly interpret all combinations of syntax at a particular profile will be deemed compliant however it works. The first problem a decoder has is that the input is an endless bitstream which contains a huge range of parameters many of which have variable length. Unique synchronizing patterns must be placed periodically throughout the bitstream so that the decoder can identify known starting points. The pictures which can be sent under MPEG-2 are so flexible that the decoder must first find a sequence header so that it can establish the size of the picture, the frame rate, the colour coding used, etc.

The decoder must also be supplied with a 27 MHz system clock. In a DVD player, this would come from a crystal, but in a transmission system this would be provided by a numerically locked loop running from the clock reference parameter in the bitstream (see Chapter 6). Until this loop has achieved lock the decoder cannot function properly.

Figure 5.41 shows a bidirectional decoder. The decoder can only begin decoding with an *I* picture and as this only uses intra-coding there will be no vectors. An *I* picture is transmitted as a series of slices. These slices begin with subsidiary synchronizing patterns. The first macroblock in the slice contains an absolute DC coefficient, but the remaining macroblocks code the DC coefficient differentially so the decoder must subtract the differential values from the previous value to obtain the absolute value.

The AC coefficients are sent as Huffman coded run/size parameters followed by coefficient value codes. The variable-length Huffman codes are decoded by using a look-up table and extending the number of bits considered until a match is obtained. This allows the zero-run-length and the coefficient size to be established. The right number of bits is taken from the bitstream corresponding to the coefficient code and this is decoded to the actual coefficient using the size parameter.

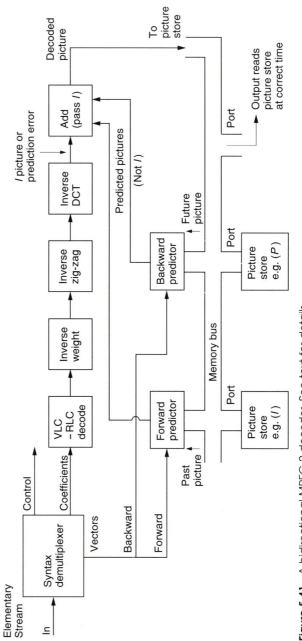

Figure 5.41 A bidirectional MPEG-2 decoder. See text for details.

If the correct number of bits has been taken from the stream, the next bit must be the beginning of the next run/size code and so on until the EOB (end of block) symbol is reached. The decoder uses the coefficient values and the zero-run-lengths to populate a DCT coefficient block following the appropriate zig-zag scanning sequence. Following EOB, the bitstream then continues with the next DCT block. Clearly this Huffman decoding will work perfectly or not at all. A single-bit slippage in synchronism or a single corrupted data bit can cause a spectacular failure.

Once a complete DCT coefficient block has been received, the coefficients need to be inverse quantized and inverse weighted. Then an inverse DCT can be performed and this will result in an 8×8 pixel block. A series of DCT blocks will allow the luminance and colour information for an entire macroblock to be decoded and this can be placed in a frame store. Decoding continues in this way until the end of the slice when an absolute DC coefficient will once again be sent. Once all the slices have been decoded, an entire picture will be resident in the frame store.

The amount of data needed to decode the picture is variable and the decoder just keeps going until the last macroblock is found. It will obtain data from the input buffer. In a constant bit rate transmission system, the decoder will remove more data to decode an *I* picture than has been received in one picture period, leaving the buffer emptier than it began. Subsequent *P* and *B* pictures need much less data and allow the buffer to fill again. The picture will be output when the timestamp (see Chapter 6) sent with the picture matches the state of the decoder's time count.

Following the *I* picture may be another *I* picture or a *P* picture. Assuming a *P* picture, this will be predictively coded from the *I* picture. The *P* picture will be divided into slices as before. The first vector in a slice is absolute, but subsequent vectors are sent differentially. However, the DC coefficients are not differential.

Each macroblock may contain a forward vector. The decoder uses this to shift pixels from the *I* picture into the correct position for the predicted *P* picture. The vectors have half-pixel resolution and where a half-pixel shift is required, an interpolator will be used.

The DCT data is sent much as for an *I* picture, it will require inverse quantizing, but not inverse weighting because *P* and *B* coefficients are flat-weighted. When decoded this represents an error cancelling picture which is added pixel-by-pixel to the motion-predicted picture. This results in the output picture.

If bidirectional coding is being used, the *P* picture may be stored until one or more *B* pictures have been decoded. The *B* pictures are sent essentially as a *P* picture might be, except that the vectors can be forward, backward or bidirectional. The decoder must take pixels from the *I* picture, the *P* picture, or both, and shift them according to the vectors to make a predicted picture. The DCT data decodes to produce an error cancelling image as before.

In an interlaced system, the prediction mechanism may alternatively obtain pixel data from the previous field or the field before that. Vectors may relate to macroblocks or to 16×8 pixel areas. DCT blocks after decoding may represent frame lines or field lines. This adds up to a lot of different possibilities for a decoder handling an interlaced input.

5.19   Coding artifacts

This section describes the visible results of imperfect coding. Imperfect coding may be where the coding algorithm is sub-optimal, where the coder latency is too short or where the compression factor in use is simply too great for the material.

In motion-compensated systems such as MPEG, the use of periodic intra-fields means that the coding noise varies from picture to picture and this may be visible as noise pumping. Noise pumping may also be visible where the amount of motion changes. If a pan is observed, as the pan speed increases the motion vectors may become less accurate and reduce the quality of the prediction processes. The prediction errors will get larger and will have to be more coarsely quantized. Thus the picture gets noisier as the pan accelerates and the noise reduces as the pan slows down. The same result may be apparent at the edges of a picture during zooming. The problem is worse if the picture contains fine detail. Panning on grass or trees waving in the wind taxes most coders severely. Camera-shake from a hand-held camera also increases the motion vector data and results in more noise as does film weave.

Input video noise or film grain degrades inter-coding as there is less redundancy between pictures and the difference data become larger, requiring coarse quantizing and adding to the existing noise.

Where a codec is really fighting the quantizing may become very coarse and as a result the video level at the edge of one DCT block may not match that of its neighbour. As a result the DCT block structure becomes visible as a mosaicing or tiling effect. Coarse quantizing also causes some coefficients to be rounded up and appear larger than they should be. High-frequency coefficients may be eliminated by heavy quantizing and this forces the DCT to act as a steep-cut low-pass filter. This causes fringeing or ringing around sharp edges and extra shadowy edges which were not in the original. This is most noticeable on text.

Excess compression may also result in colour bleed where fringeing has taken place in the chroma or where high-frequency chroma coefficients have been discarded. Graduated colour areas may reveal banding or posterizing as the colour range is restricted by requantizing. These artifacts are almost impossible to measure with conventional test gear.

Neither noise pumping nor blocking are visible on analog video recorders and so it is nonsense to liken the performance of a codec to the quality of a VCR. In fact noise pumping is extremely objectionable because, unlike steady noise, it attracts attention in peripheral vision and may result in viewing fatigue.

In addition to highly detailed pictures with complex motion, certain types of video signal are difficult for MPEG-2 to handle and will usually result in a higher level of artifacts than usual. Noise has already been mentioned as a source of problems. Timebase error from, for example, VCRs is undesirable because this puts succesive lines in different horizontal positions. A straight vertical line becomes jagged and this results in high spatial frequencies in the DCT process. Spurious coefficients are created which need to be coded.

Much archive video is in composite form and MPEG-2 can only handle this after it has been decoded to components. Unfortunately many general-purpose composite decoders have a high level of residual sub-carrier in the outputs. This is normally not a problem because the sub-carrier is designed to be invisible to the naked eye. Figure 5.42 shows that in PAL and NTSC the sub-carrier frequency is selected so that a phase reversal is achieved between successive lines and frames.

While this makes the sub-carrier invisible to the eye, it is not invisible to an MPEG decoder. The sub-carrier waveform is interpreted as a horizontal frequency, the vertical phase reversals are interpreted as a vertical spatial frequency and the picture-to-picture reversals increase the

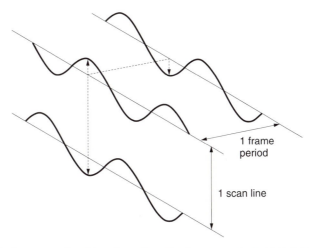

Figure 5.42 In composite video the subcarrier frequency is arranged so that inversions occur between adjacent lines and pictures to help reduce the visibility of the chroma.

magnitude of the prediction errors. The sub-carrier level may be low but it can be present over the whole screen and require an excess of coefficients to describe it.

Composite video should not in general be used as a source for MPEG-2 encoding, but where this is inevitable the standard of the decoder must be much higher than average, especially in the residual sub-carrier specification. Some MPEG preprocessors support high-grade composite decoding options.

Judder from conventional linear standards convertors degrades the performance of MPEG-2. The optic flow axis is corrupted and linear filtering causes multiple images which confuse motion estimators and result in larger prediction errors. If standards conversion is necessary, the MPEG-2 system must be used to encode the signal in its original format and the standards convertor should be installed after the decoder. If a standards convertor has to be used before the encoder, then it must be a type which has effective motion compensation.

Film weave causes movement of one picture with respect to the next and this results in more vector activity and larger prediction errors. Movement of the centre of the film frame along the optical axis causes magnification changes which also result in excess prediction error data. Film grain has the same effect as noise: it is random and so cannot be compressed.

Perhaps because it is relatively uncommon, MPEG-2 cannot handle image rotation well because the motion-compensation system is only designed for translational motion. Where a rotating object is highly detailed, such as in certain fairground rides, the motion-compensation failure requires a significant amount of prediction error data and if a suitable bit rate is not available the level of artifacts will rise.

Flash guns used by still photographers are a serious hazard to MPEG-2 especially when long GOPs are used. At a press conference where a series of flashes may occur, the resultant video contains intermittent white frames which defeat prediction. A huge prediction error is required to turn the previous picture into a white picture, followed by another huge prediction error to return the white frame to the next picture. The output buffer fills and heavy requantizing is employed. After a few flashes the picture has generally gone to tiles.

5.20   Processing MPEG-2 and concatenation

Concatenation loss occurs when the losses introduced by one codec are compounded by a second codec. All practical compressers, MPEG-2 included, are lossy because what comes out of the decoder is not bit-identical to what went into the encoder. The bit differences are controlled so that they have minimum visibility to a human viewer.

MPEG-2 is a toolbox which allows a variety of manipulations to be performed in both the spatial and the temporal domain. There is a limit to the compression which can be used on a single frame, and if higher compression factors are needed, temporal coding will have to be used. The longer the run of pictures considered, the lower the bit rate needed, but the harder it becomes to edit.

The most editable form of MPEG-2 is to use *I* pictures only. As there is no temporal coding, pure cut edits can be made between pictures. The next best thing is to use a repeating *IB* structure which is locked to the odd/even field structure. Cut edits cannot be made as the *B* pictures are bidirectionally coded and need data from both adjacent *I* pictures for decoding. The *B* picture has to be decoded prior to the edit and re-encoded after the edit. This will cause a small concatenation loss.

Beyond the *IB* structure processing gets harder. If a long GOP is used for the best compression factor, an *IBBPBBP . . .* structure results. Editing this is very difficult because the pictures are sent out of order so that bidirectional decoding can be used. MPEG allows closed GOPs where the last *B* picture is coded wholly from the previous pictures and does not need the *I* picture in the next GOP. The bitstream can be switched at this point but only if the GOP structures in the two source video signals are synchronized (makes colour framing seem easy). Consequently in practice a long GOP bitstream will need to be decoded prior to any production step. Afterwards it will need to be re-encoded.

This is known as *naive* concatenation and an enormous pitfall awaits. Unless the GOP structure of the output is identical to and synchronized with the input the results will be disappointing. The worst case is where an *I* picture is encoded from a picture which was formerly a *B* picture. It is easy enough to lock the GOP structure of a coder to a single input, but if an edit is made between two inputs, the GOP timings could well be different.

As there are so many structures allowed in MPEG, there will be a need to convert between them. If this has to be done, it should only be in the direction which increases the GOP length and reduces the bit rate. Going the other way is inadvisable. The ideal way of converting from, say, the *IB* structure of a news system to the *IBBP* structure of an emission system is to use a recompressor. This is a kind of standards convertor which will give better results than a decode followed by an encode.

The DCT part of MPEG-2 itself is lossless. If all the coefficients are preserved intact an inverse transform yields the same pixel data. Unfortunately this does not yield enough compression for many applications. In practice the coefficients are made less accurate by removing bits, starting at the least significant end and working upwards. This process is weighted, or made progressively more aggressive as spatial frequency increases. Small-value coefficients may be truncated to zero and large-value coefficients are most coarsely truncated at high spatial frequencies where the effect is least visible.

Figure 5.43(a) shows what happens in the ideal case where two *identical* coders are put in tandem and synchronized. The first coder quantizes the coefficients to finite accuracy and causes a loss on decoding. However, when the second coder performs the DCT calculation, the coefficients obtained will be identical to the quantized coefficients in the first coder and so if the second weighting and requantizing step is identical the same truncated coefficient data will result and there will be no further loss of quality.[3]

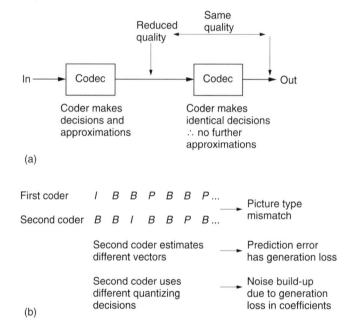

(a)

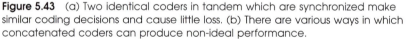

(b)

Figure 5.43 (a) Two identical coders in tandem which are synchronized make similar coding decisions and cause little loss. (b) There are various ways in which concatenated coders can produce non-ideal performance.

In practice this ideal situation is elusive. If the two DCTs become non-identical for any reason, the second requantizing step will introduce further error in the coefficients and the artifact level goes up. Figure 5.43(b) shows that non-identical concatenation can result from a large number of real-world effects.

An intermediate processing step such as a fade will change the pixel values and thereby the coefficients. A DVE resize or shift will move pixels from one DCT block to another. Even if there is no processing step, this effect will also occur if the two codecs disagree on where the MPEG picture boundaries are within the picture. If the boundaries are correct there will still be concatenation loss if the two codecs use different weighting.

One problem with MPEG is that the compressor design is unspecified. While this has advantages, it does mean that the chances of finding identical coders is minute because each manufacturer will have his own views on the best compression algorithm. In a large system it may be worth obtaining the coders from a single supplier.

It is now increasingly accepted that concatenation of compression techniques is potentially damaging, and results are worse if the codecs are different. Clearly feeding a digital coder such as MPEG-2 with a signal which has been subject to analog compression comes into the category of worse. Using interlaced video as a source for MPEG coding is sub-optimal and using decoded composite video is even worse.

One way of avoiding concatenation is to stay in the compressed data domain. If the goal is just to move pictures from one place to another, decoding to traditional video so an existing router can be used is not ideal, although substantially better than going through the analog domain.

Figure 5.44 shows some possibilities for picture transport. Clearly if the pictures exist as a compressed file on a server, a file transfer is the right way to do it as there is no possibility of loss because there has been no

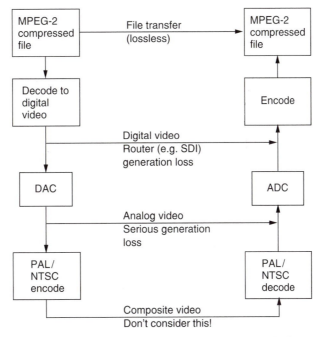

Figure 5.44 Compressed picture transport mechanisms contrasted.

concatenation. File transfer is also quite indifferent to the picture format. It doesn't care whether the pictures are interlaced or not, whether the colour is 4:2:0 or 4:2:2.

Decoding to SDI (serial digital interface) standard is sometimes done so that existing serial digital routing can be used. This is concatenation and has to be done carefully. The compressed video can only use interlace with non-square pixels and the colour coding has to be 4:2:2 because SDI only allows that. If a compressed file has 4:2:0 the chroma has to be interpolated up to 4:2:2 for SDI transfer and then sub-sampled back to 4:2:0 at the second coder and this will cause generation loss. An SDI transfer also can only be performed in real time, thus negating one of the advantages of compression. In short, traditional SDI is not really at home with compression.

As 4:2:0 progressive scan gains popularity and video production moves steadily towards non-format-specific hardware using computers and data networks, use of the serial digital interface will eventually decline. In the short term, if an existing SDI router has to be used, one solution is to produce a bitstream which is sufficiently similar to SDI that a router will pass it. In other words the signal level, frequency and impedance is pure SDI, but the data protocol is different so that a bit-accurate file transfer can be performed. This has two advantages over SDI. First, the compressed data format can be anything appropriate and non-interlaced and/or 4:2:0 can be handled in any picture size, aspect ratio or frame rate. Second, a faster than real-time transfer can be used depending on the compression factor of the file. Equipment which allows this is becoming available and its use can mean that the full economic life of a SDI routing installation can be obtained.

An improved way of reducing concatenation loss has emerged from the ATLANTIC research project.[4] Figure 5.45 shows that the second encoder in a concatenated scheme does not make its own decisions from the incoming video, but is instead steered by information from the first bitstream. As the second encoder has less intelligence, it is known as a *dim* encoder.

The information bus carries all the structure of the original MPEG-2 bitstream which would be lost in a conventional decoder. The ATLANTIC

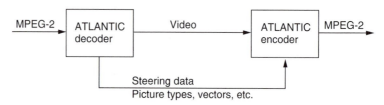
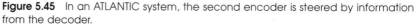

Figure 5.45 In an ATLANTIC system, the second encoder is steered by information from the decoder.

decoder does more than decode the pictures. It also places on the information bus all parameters needed to make the dim encoder re-enact what the initial MPEG-2 encode did as closely as possible.

The GOP structure is passed on so that pictures are re-encoded as the same type. Positions of macroblock boundaries become identical so that DCT blocks contain the same pixels and motion vectors relate to the same screen data. The weighting and quantizing tables are passed so that coefficient truncation is identical. Motion vectors from the original bitsream are passed on so that the dim encoder does not need to perform motion estimation. In this way predicted pictures will be identical to the original prediction and the prediction error data will be the same.

One application of this approach is in recompression, where an MPEG-2 bitstream has to have its bit rate reduced. This has to be done by heavier requantizing of coefficients, but if as many other parameters as possible can be kept the same, such as motion vectors, the degradation will be minimized. In a simple recompressor just requantizing the coefficients means that the predictive coding will be impaired. In a proper encode, the quantizing error due to coding, say, an I picture is removed from the P picture by the prediction process. The prediction error of P is obtained by subtracting the decoded I picture rather than the original I picture.

In simple recompression this does not happen and there may be a tolerance build-up known as drift.[5] A more sophisticated recompressor will need to repeat the prediction process using the decoded output pictures as the prediction reference.

MPEG-2 bitstreams will often be decoded for the purpose of switching. Local insertion of commercial breaks into a centrally originated bitstream is one obvious requirement. If the decoded video signal is switched, the information bus must also be switched. At the switch point identical re-encoding becomes impossible because prior pictures required for predictive coding will have disappeared. At this point the dim encoder has to become bright again because it has to create an MPEG-2 bitstream without assistance.

It is possible to encode the information bus into a form which allows it to be invisibly carried in the serial digital interface. Where a production process such as a vision mixer or DVE performs no manipulation, i.e. becomes bit transparent, the subsequent encoder can extract the information bus and operate in 'dim' mode. Where a manipulation is performed, the information bus signal will be corrupted and the encoder has to work in 'bright' mode. The encoded information signal is known as a 'mole'[6] because it burrows through the processing equipment!

There will be a generation loss at the switch point because the re-encode will be making different decisions in bright mode. This may be difficult to detect because the human visual system is slow to react to a vision cut and defects in the first few pictures after a cut are masked.

In addition to the video computation required to perform a cut, the process has to consider the buffer occupancy of the decoder. A downstream decoder has finite buffer memory, and individual encoders model the decoder buffer occupancy to ensure that it neither overflows nor underflows. At any instant the decoder buffer can be nearly full or nearly empty without a problem provided there is a subsequent correction. An encoder which is approaching a complex *I* picture may run down the buffer so it can send a lot of data to describe that picture. Figure 5.46(a) shows that if a decoder with a nearly full buffer is suddenly switched to an encoder which has been running down its buffer occupancy, the decoder buffer will overflow when the second encoder sends a lot of data.

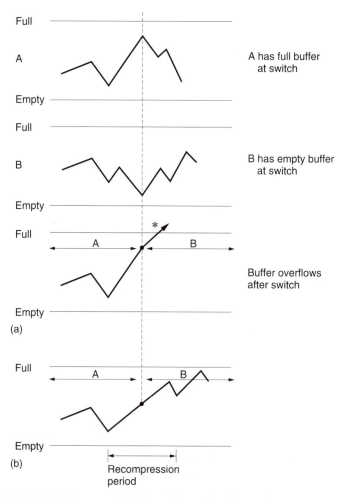

Figure 5.46 (a) A bitstream switch at a different level of buffer occupancy can cause a decoder overflow. (b) Recompression after a switch to return to correct buffer occupancy.

An MPEG-2 switcher will need to monitor the buffer occupancy of its own output to avoid overflow of downstream decoders. Where this is a possibility the second encoder will have to recompress to reduce the output bit rate temporarily. In practice there will be a recovery period where the buffer occupancy of the newly selected signal is matched to that of the previous signal. This is shown in Figure 5.46(b).

References

1. Kelly, D.H., Visual processing of moving stimuli. *J. Opt. Soc. America.* **2**, 216–225 (1985)
2. Uyttendaele, A., Observations on scanning formats. Presented at HDTV '97 Montreux (June 1997)
3. Stone, J. and Wilkinson, J., Concatenation of video compression systems. Presented at 137th SMPTE Tech. Conf., New Orleans (1995)
4. Wells, N.D., The ATLANTIC project: Models for programme production and distribution. *Proc. Euro. Conf. Multimedia Applications Services and Techniques* (ECMAST), 243–253 (1996)
5. Werner, O., Drift analysis and drift reduction for multiresolution hybrid video coding. *Image Communication*, **8**, 387–409 (1996)
6. Knee, M.J. and Wells, N.D., Seamless concatenation – a 21st century dream. Presented at Int. Television. Symp. Montreux (1997)

6

Program and transport streams

After compression, audio and video signals are simply data and only differ from generic data in that they relate to real-time signals. MPEG bitstreams are basically means of transporting data while allowing the time axis of the original signal to be recreated.

6.1   Introduction

There are two basic applications of MPEG bitstreams: recording and transmission and these have quite different requirements. In the multichannel recording application the encoders and decoders can all share the same clock. The transport speed can be altered so that the incoming bit rate can be slaved to a reference generated by the decoder itself.

In the case of multichannel transmission the program sources are not necessarily synchronous. The source of timing is the encoder and the decoder must synchronize or genlock to that. Thus the difference between a program stream and a transport stream is that the latter must contain additional synchronizing information to lock the encoder and decoder clocks together independently in each program.

6.2   Packets and time stamps

The video Elementary Stream is an endless bitstream representing pictures which are not necessarily in the correct order and which take a variable length of time to transmit. Storage and transmission systems prefer discrete blocks of data and so Elementary Streams are packetized to form a PES (packetized Elementary Stream). Audio Elementary

Streams are also packetized. A packet is shown in Figure 6.1. It begins with a header containing an unique packet start code and a code which identifies the type of data stream. Optionally the packet header also may contain one or more *time stamps* which are used for synchronizing the video decoder to real time and for obtaining lip-sync.

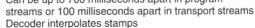

PTS/DTS need not be in every PES packet.
Can be up to 700 milliseconds apart in program
streams or 100 milliseconds apart in transport streams
Decoder interpolates stamps

Figure 6.1 A PES packet structure is used to break up the continuous Elementary Stream.

Figure 6.2 shows that a time stamp is a sample of the state of a counter which is driven by a 90 KHz clock. This is obtained by dividing down the master 27 MHz clock of MPEG-2. There are two types of time stamp: PTS and DTS. These are abbreviations for presentation time stamp and decode time stamp. A presentation time stamp determines when the associated picture should be displayed on the screen, whereas a decode time stamp determines when it should be decoded. In bidirectional coding these times can be quite different.

Figure 6.2 Time stamps are the result of sampling a counter driven by the encoder clock.

Audio packets only have presentation time stamps. Clearly if lip-sync is to be obtained, the audio and the video streams of a given program must have been locked to the same master 27 MHz clock and the time stamps must have come from the same counter driven by that clock.

With reference to Figure 6.3, the GOP begins with an *I* picture, and then *P1* is sent out of sequence prior to the *B* pictures. *P1* has to be decoded before *B1* and *B2* can be decoded. As only one picture can be decoded at a time, the *I* picture is decoded at time *N*, but not displayed until time *N*

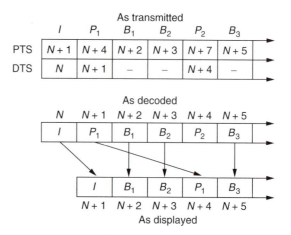

Figure 6.3 An example of using PTS/DTS to synchronize bidirectional decoding.

+ 1. As the *I* picture is being displayed, *P1* is being decoded at *N* + 1. *P1* will be stored in RAM. At time *N* + 2, *B1* is decoded and displayed immediately. For this reason *B* pictures need only PTS. At *N* + 3, *B2* is decoded and displayed. At *N* + 4, *P1* is displayed, hence the large difference between PTS and DTS in P1. Simultaneously *P2* is decoded and stored ready for the decoding of *B3* and so on.

In practice the time between input pictures is constant and so there is a certain amount of redundancy in the time stamps. Consequently PTS/DTS need not appear in every PES packet. Time stamps can be up to 700 milliseconds apart in program streams and up to 100 milliseconds apart in transport streams. As each picture type (*I, P* or *B*) is flagged in the bitstream, the decoder can infer the PTS/DTS for every picture from the ones actually transmitted.

Figure 6.4 shows that one or more PES packets can be assembled in a *pack* whose header contains a sync pattern and a *system clock reference* code which allows the decoder to recreate the encoder clock. Clock reference operation is described in section 6.4.

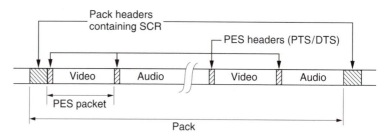

Figure 6.4 A pack is a set of PES packets. The pack header contains a clock reference code.

6.3 Transport streams

The MPEG-2 transport stream is intended to be a multiplex of many TV programs with their associated sound and data channels, although a single program transport stream (SPTS) is possible. The transport stream is based upon packets of constant size so that adding error correction codes and interleaving[1] in a higher layer is eased. Figure 6.5 shows that these are always 188 bytes long. Transport stream packets should not be confused with PES packets which are larger and which vary in size.

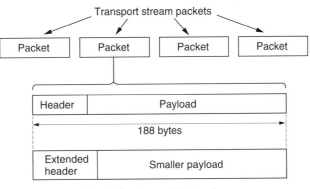

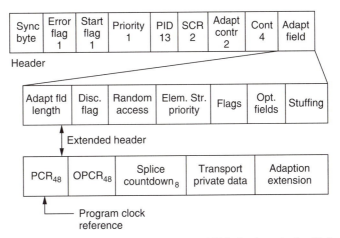

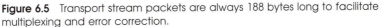

Figure 6.5 Transport stream packets are always 188 bytes long to facilitate multiplexing and error correction.

Transport stream packets always begin with a header. The remainder of the packet carries data known as the payload. For efficiency, the normal header is relatively small, but for special purposes the header may be extended. In this case the payload gets smaller so that the overall size of the packet is unchanged.

The header begins with a sync byte which is an unique pattern detected by a demultiplexer. A transport stream may contain many different elementary streams and these are identified by giving each an unique 13-bit Packet Identification Code or PID which is included in the header. A multiplexer seeking a particular elementary stream simply checks the PID of every packet and accepts those which match, rejecting the rest.

In a multiplex there may be many packets from other programs in between packets of a given PID. To help the demultiplexer, the packet header contains a continuity count. This is a 4-bit value which increments at each new packet having a given PID.

This approach allows statistical multiplexing as it does matter how many or how few packets have a given PID; the demux will still find them. Statistical multiplexing has the problem that it is virtually impossible to make the sum of the input bit rates constant. Instead the multiplexer aims to make the average data bit rate slightly less than the maximum and the overall bit rate is kept constant by adding 'stuffing' or null packets. These packets have no meaning, but simply keep the bit rate constant. Null packets always have a PID of 8191 (all ones) and the demultiplexer discards them.

6.4   Clock references

A transport stream is a multiplex of several TV programs and these may have originated from widely different locations. It is impractical to expect all the programs in a transport stream to be synchronous and so the stream is designed from the outset to allow asynchronous programs. A decoder running from a transport stream has to genlock to the encoder and the transport stream has to have a mechanism to allow this to be done independently for each program. The synchronizing mechanism is called Program Clock Reference (PCR).

In program streams all the programs must be synchronous so that only one clock is required at the decoder. In this case the synchronizing mechanism is called System Clock Reference (SCR).

Figure 6.6 shows how the PCR/SCR system works. The goal is to recreate at the decoder a 27 MHz clock which is synchronous with that at the encoder. The encoder clock drives a 48-bit counter which continuously counts up to the maximum value before overflowing and beginning again.

A transport stream multiplexer will periodically sample the counter and place the state of the count in an extended packet header as a PCR (see Figure 6.5). The demultiplexer selects only the PIDs of the required program, and it will extract the PCRs from the packets in which they were inserted. In a program stream the count is placed in a pack header as an SCR which the decoder can identify.

The PCR/SCR codes are used to control a numerically locked loop (NLL). This is similar to a phase-locked loop, except that the two phases concerned are represented by the state of a binary number. The NLL contains a 27 MHz VCXO (Voltage Controlled Crystal Oscillator), a variable-frequency oscillator based on a crystal which has a relatively small frequency range.

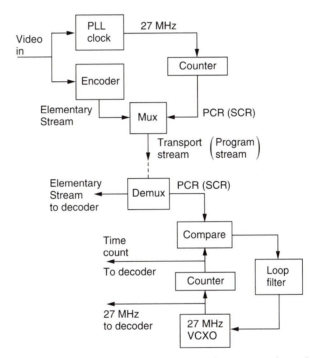

Figure 6.6 Program or System Clock Reference codes regenerate a clock at the decoder. See text for details.

The VCXO drives a 48-bit counter in the same way as in the encoder. The state of the counter is compared with the contents of the PCR/SCR and the difference is used to modify the VCXO frequency. When the loop reaches lock, the decoder counter would arrive at the same value as is contained in the PCR/SCR and no change in the VCXO would then occur. In practice the transport stream packets will suffer from transmission jitter and this will create phase noise in the loop. This is removed by the loop filter so that the VCXO effectively averages a large number of phase errors.

A heavily damped loop will reject jitter well, but will take a long time to lock. Lock-up time can be reduced when switching to a new program if the decoder counter is jammed to the value of the first PCR received in the new program. The loop filter may also have its time constants shortened during lock-up. Once a synchronous 27 MHz clock is available at the decoder, this can be divided down to provide the 90 KHz clock which drives the time stamp mechanism. The entire timebase stability of the decoder is no better than the stability of the clock derived from PCR/SCR. MPEG-2 sets standards for the maximum amount of jitter which can be present in PCRs in a real transport stream.

6.5   Program Specific Information (PSI)

In a real transport stream, each Elementary Stream has a different PID, but the demultiplexer has to be told what these PIDs are and what audio belongs with what video before it can operate. This is the function of PSI. Figure 6.7 shows the structure of PSI. When a decoder powers up, it knows nothing about the incoming transport stream except that it must search for all packets with a PID of zero. PID zero is reserved for the Program Association Table (PAT). The PAT is transmitted at regular intervals and contains a list of all the programs in this transport stream. Each program is further described by its own Program Map Table (PMT) and the PIDs of of the PMTs are contained in the PAT.

Figure 6.7 also shows that the PMTs fully describe each program. The PID of the video Elementary Stream is defined, along with the PID(s) of

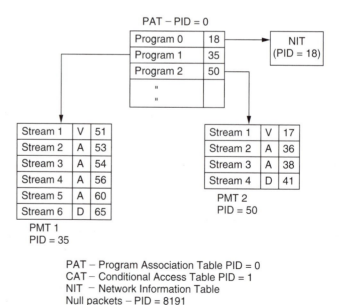

Figure 6.7   MPEG-2 Program Specific Information (PSI) is used to tell a demultiplexer what the transport stream contains.

the associated audio and data streams. Consequently when the viewer selects a particular program, the demultiplexer looks up the program number in the PAT, finds the right PMT and reads the audio, video and data PIDs. It then selects Elementary Streams having these PIDs from the transport stream and routes them to the decoders.

Program 0 of the PAT contains the PID of the Network Information Table (NIT). This contains information about what other transport streams are available. For example, in the case of a satellite broadcast, the NIT would detail the orbital position, the polarization, carrier frquency and modulation scheme. Using the NIT a set top box could automatically switch between transport streams.

Apart from 0 and 8191, a PID of 1 is also reserved for the Conditional Access Table (CAT). This is part of the access control mechanism needed to support pay per view or subscription viewing.

6.6   Multiplexing

A transport stream multiplexer is a complex device because of the number of functions it must perform. A fixed multiplexer will be considered first. In a fixed multiplexer, the bit rate of each of the programs must be specified so that the sum does not exceed the payload bit rate of the transport stream. The payload bit rate is the overall bit rate less the packet headers and PSI rate.

In practice the programs will not be synchronous to one another, but the transport stream must produce a constant packet rate given by the bit rate divided by 188 bytes, the packet length. Figure 6.8 shows how this is

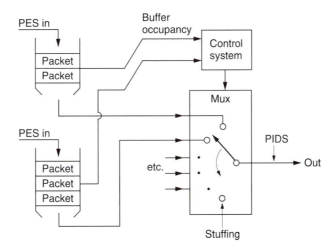

Figure 6.8 A transport stream multiplexer can handle several programs which are asynchronous to one another and to the transport stream clock. See text for details.

handled. Each elementary stream entering the multiplexer passes through a buffer which is divided into payload-sized areas. Note that periodically the payload area is made smaller because of the requirement to insert PCR.

MPEG-2 decoders also have a quantity of buffer memory. The challenge to the multiplexer is to take packets from each program in such a way that neither its own buffers nor the buffers in any decoder either overflow or underflow. This requirement is met by sending packets from all programs as evenly as possible rather than bunching together a lot of packets from one program. When the bit rates of the programs are different, the only way this can be handled is to use the buffer contents indicators. The fuller a buffer is, the more likely it should be that a packet will be read from it. This a buffer content arbitrator can decide which program should have a packet allocated next.

If the sum of the input bit rates is correct, the buffers should all slowly empty because the overall input bit rate has to be less than the payload bit rate. This allows for the insertion of Program Specific Information. While PATs and PMTs are being transmitted, the program buffers will fill up again. The multiplexer can also fill the buffers by sending more PCRs as this reduces the payload of each packet. In the event that the multiplexer has sent enough of everything but still can't fill a packet then it will send a null packet with a PID of 8191. Decoders will discard null packets and as they convey no useful data, the multiplexer buffers will all fill while null packets are being transmitted.

In a statistical multiplexer or statmux, the bit rate allocated to each program can vary dynamically. Figure 6.9 shows that there must be tight connection between the statmux and the associated compressors. Each compressor has a buffer memory which is emptied by a demand clock from the statmux. In a normal, fixed bit rate, coder the buffer contents feeds back and controls the requantizer. In statmuxing this process is less

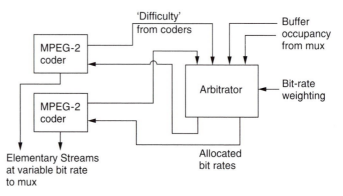

Figure 6.9 A statistical multiplexer contains an arbitrator which allocates bit rate to each program as a function of program difficulty.

severe and only takes place if the buffer is very close to full, because the buffer content is also fed to the statmux.

The statmux contains an arbitrator which allocates packets to the program with the fullest buffer. Thus if a particular program encounters difficult material it will produce large prediction errors and begin to fill its output buffer. This will cause the statmux to allocate more packets to that program. In order to fill more packets, the statmux clocks more data out of that buffer, causing the level to fall again. Of course, this is only possible if the other programs in the transport stream are handling typical video.

In the event that several programs encounter difficult material at once, clearly the buffer contents will rise and the requantizing mechanism will have to operate.

6.7   Remultiplexing

In real life a program creator may produce a transport stream which carries all its programs simultaneously. A service provider may take in several such streams and create its own transport stream by selecting different programs from different sources. In an MPEG-2 environment this requires a remultiplexer, also known as a transmultiplexer. Figure 6.10 shows what a remultiplexer does.

Remultiplexing is easier when all the incoming programs have the same bit rate. If a suitable combination of programs is selected it is obvious that the output transport stream will always have sufficient bit rate. Where statistical multiplexing has been used, there is a possibility that the sum of the bit rates of the selected programs will exceed the bit

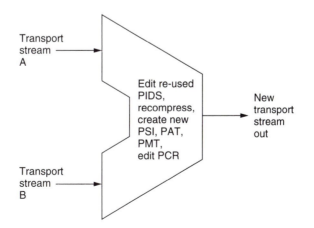

Figure 6.10 A remultiplexer creates a new transport stream from selected programs in other transport streams.

rate of the output transport stream. To avoid this, the remultiplexer will have to employ recompression.

Recompression reqires a partial decode of the bitstream to identify the DCT coefficients. These will then be requantized to reduce the bit rate until it is low enough to fit the output transport stream.

Remultiplexers have to edit the Program Specific Information (PSI) such that the Program Association Table (PAT) and the Program Map Tables (PMT) correctly reflect the new transport stream content. It may also be necessary to change the packet identification codes (PIDs) since the incoming transport streams could have inadvertently used the same values.

When Program Clock Reference (PCR) data are included in an extended packet header, they represent a real-time clock count and if the associated packet is moved in time the PCR value will be wrong. Remultiplexers have to recreate a new multiplex from a number of other multiplexes and it is inevitable that this process will result in packets being placed in different locations in the output transport stream than they had in the input. In this case the remultiplexer must edit the PCR values so that they reflect the value the clock counter would have had at the location the packet at which the packet now resides.

Reference

1. Watkinson, J.R., *The Art of Digital Video*, Chapter 6, Oxford: Focal Press (1994)

Glossary

AAU Audio access unit (*see* Access unit).

AC-3 Audio coding technique developed by Dolby used with ATSC (q.v.) and DVD (q.v.) aka Dolby Digital.

ACATS Advisory Committee on Advanced Television Service.

Access unit The coded data for a picture or block of sound and any stuffing which follows it.

Anchor picture Picture used as basis for predictive coding.

ATM Asynchronous transfer mode – a protocol for transport in broadband digital networks.

ATSC Advanced Television Systems Committee.

Bouquet Group of transport streams in which programs are identified by combination of network ID and PID. Part of DVB-SI.

CAT Conditional access table. Packets having PID (q.v.) code of 1 which contain information about the scrambling system (*see* ECM and EMM).

Channel code Modulation technique which converts raw data into a signal which can be recorded, or transmitted by radio or cable.

CIF Common interchange format. 352 $\times$ 240 pixel format for 30 fps videoconferencing.

Closed GOP Group of pictures in which the last pictures do not need data from the next GOP for bidirectional coding. Used to make a splice point in a bitstream.

Coefficient number specifying the amplitude of a particular frequency in a transform.

Concatenation Connecting more than one codec in tandem.

DCT Discrete cosine transform.

DTS Decoding time stamp. Part of PES header indicating when an access unit is to be decoded.

DVB Digital video broadcasting. The broadcasting of television programs using digital modulation of a radio frequency carrier. This can be terrestrial or from a satellite.

DVB-IRD *See* IRD.

DVB-SI DVB Service information. Information carried in a DVB multiplex describing the contents of different multiplexes. Includes NIT, SDT, EIT, TDT, BAT, RST, ST (q.v.).

DVC Digital video cassette.

DVD Digital video disk, aka digital versatile disk. Optical disk for consumer video or data applications.

Differential coding Sending a value not in an absolute sense but as the difference between the current and previous values.

Drift Error build-up when bit rate is further reduced by additional truncation of coefficients.

ECM Entitlement control message. Conditional access information specifying control words or other stream-specific scrambling parameters.

EIT Event information table. Part of DVB-SI.

EMM Entitlement management message. Conditional access information specifying authorization level or services of specific decoders. An individual decoder or a group of decoders may be addressed.

ENG Electronic news gathering. Term used to describe use of video-recording instead of film in news coverage.

EOB End of block – code sent when all remaining coefficients in a DCT block are zero.

EPG Electronic program guide. Program guide delivered by data transfer rather than printed paper.

Elementary Stream Raw output of a compresser carrying a single video or audio signal.

Entropy coding Coding system which achieves compression of signals having non-uniform statistics.

Error propagation When a small error in a critical bit results in protracted errors in a decoder.

Event A set of Elementary Streams, typically audio and video, having common clocks, start times and end times.

FEC Forward error correction. System in which redundancy is added to the message so that errors can be corrected dynamically at the receiver.

GOP Group of pictures starts with an *I* picture and ends with last picture before next *I* picture.

Huffman coding Form of entropy coding (q.v.) using variable-length parameters.

Inter-coding Compression which uses redundancy between successive pictures, aka temporal coding.

Interleaving Technique used with error correction which breaks up burst errors into many smaller errors.

Intra-coding Compression which works entirely within one picture, aka spatial coding.

IRD Integrated receiver decoder. Combined RF receiver and MPEG decoder used to adapt TV set to digital transmissions.

Level The size of the input picture in use with a given profile (q.v.)

Macroblock Screen area represented by several luminance and colour difference DCT blocks which are all steered by one motion vector.

Masking Psycho-acoustic phenomenon whereby certain sounds cannot be heard in the presence of others.

Mezzanine level System using lower compression factor to permit concatenation.

NIT Network information table. Information in one transport stream which describes many transport streams.

Null packets Packets of 'stuffing' which carry no data but which are necessary to maintain a constant bit rate with a variable payload. Null packets always have a PID (q.v.) of 8191 (all 1s).

Overflow When a buffer memory consistently receives more data than is being removed.

PAT Program Association Table. Data appearing in packets having PID (q.v.) code of zero which the MPEG decoder uses to determine which programs exist in a transport stream. PAT points to PMT (q.v.) which in turn points to the video, audio and data content of each program.

PCM Pulse code modulation. Technical term for analog source waveform, e.g. audio or video, expressed as periodic numerical samples. PCM is an uncompressed digital signal.

PCR Program clock reference. Sample of encoder clock count sent in program header to synchronize decoder clock.

PCRI Interpolated program clock reference. PCR estimated from previous PCR to measure jitter.

PID Packet identifier: 13-bit code in transport packet header. PID 0 indicates packet contains PAT (q.v.). PID 1 indicates packet contains CAT (q.v.). PID 8191 (all 1s) indicats null (stuffing) packets. All packets belonging to the same elementary stream have the same PID.

PMT Program Map Tables. Tables in PAT (q.v.) which point to video, audio and data content of a transport stream.

PSI Program Specific Information. Information which keeps track of the different programs in an MPEG transport stream and the elementary streams in each program. PSI includes PAT, PMT, NIT, CAT, ECM and EMM.

PSI/SI General term for MPEG PSI and DVB-SI combined.

PTS Presentation time stamp – time at which a presentation unit is to be available to the viewer.

PU   Presentation unit. One compressed picture or block of audio.

Pack   a set of PES packets. The pack header contains a SCR code.

Packets   Beware! The term is used in two contexts and these are not the same. In program streams, a packet contains one or more presentation units. In transport streams a packet is a small fixed-size data quantum.

Padding   *See* Stuffing.

Payload   Content of a packet other than the header.

Pre-processing   Noise reduction, downsampling, cut edit identification and 3:2 pulldown identification are all pre-processing steps needed before compression.

Profile   A subset of the entire coding repertoire of MPEG.

Program stream   Bitstream containing compressed video and audio and timing information.

QCIF   One quarter resolution (176 × 144 pixels) common interchange format (*see* CIF).

QSIF   One quarter resolution source input format (*see* SIF).

RLC   Run-length coding. Coding scheme which counts number of similar bits instead of sending them individually.

SCR   System clock reference – clock data carried in program stream pack header.

STC   System time clock. Common clock used to encode video and audio in the same program.

SDT   Service description table. Table listing the providers of each service in a transport stream.

SI   *See* DVB-SI.

SIF   Source input format. Half-resolution input signal used by MPEG-1.

ST   Stuffing table.

Scalability   System where more complex decoder produces better picture but simple decoder only uses part of data.

Stuffing   Meaningless data added to maintain constant bit rate.

Syndrome   Initial result of an error-checking calculation. Generally if zero there is assumed to be no error.

TDAC   Time-domain aliasing cancellation. Coding technique used in AC-3 audio compression.

TDT   Time and date table. Used in DVB-SI.

T-STD   Transport stream system target decoder. Decoder having a certain amount of buffer memory assumed present by an encoder.

Transport stream   Multiplex of several program streams which are carried in packets. Demultiplexing is achieved by different packet IDs (PIDs) (*see* PSI, PAT, PMT, PCR).

Truncation   Shortening wordlength of sample or coefficient by removing low-order bits.

Underflow   When a buffer memory consistently receives less data than is being removed from it.

VAU Video access unit. One compressed picture in program stream.

VBV Video Buffer Verifier – Parameter specifying amount of buffer memory needed to decode a given Elementary Stream.

VLC Variable-length coding. Compression technique which allocates short codes to frequent values and long codes to infrequent values.

VOD Video-on-demand. System in which television programs or movies are transmitted to a single consumer only when required.

Vector Motion-compensation parameter which tells a decoder how to shift part of another picture to more closely approximate the current picture.

Wavelet Transform in which basis function is not of fixed length but grows longer as frequency reduces.

Weighting Method of changing distribution of noise due to truncation by pre-multiplying values.

Zero-run-length Parameter specifying the number of contiguous DCT coefficients of value zero in a scanning sequence.

Zig-zag scan Method of ordering DCT coefficients so that zero values tend to be at the end of the sequence.

Index